The "7 Churches" Of The "Revelation"

What The "Hubble" Will Never See

Sir Isaac Newton's "Plan Of The World"

Charles S. Brown

THE "7 CHURCHES" OF THE "REVELATION"

What The "Hubble" Will Never See

Sir Isaac Newton's "Plan Of The World"

Charles S. Brown

http://www.crystalbooks.org

This Edition published in **New Zealand** by
Crystal Publishing Ltd.
P.O. Box 60042, Titirangi
West Auckland
NEW ZEALAND

First Edition 2011
Copyright © 2011 Charles S. Brown

ISBN 978-0-9582813-7-9

Contents

Acknowledgements

Crystal Publishing gratefully and singularly acknowledges Ferrar Fenton – long deceased from the physical world – yet whose monumental work of re-translating The Bible finally permitted a certain key question between science and religion in centuries-old conflicts i.e., – Creation versus Evolution – to be perfectly reconciled. Through his intuitively more-correct translation of **The Book of Genesis**, particularly Chapters 1 and 2, he has singularly rendered *every other* "Genesis" translation – that does not accord with his powerfully-guided, more-correct insights – irrelevant. Near-future events will unequivocally bear out the truth of this statement.

On the matter of Creation versus Evolution, his crucial spiritual insights have now returned to **The Creator** that which is **His – the Majesty and Power of The Pure Truth of His stupendous and humanly incomprehensible Creation**.
Fenton has thus bequeathed to the worlds of science and religion the ordained foundation upon which to build, within those Disciplines:

The 'Harmonising Truth' about Creation and Evolution; and the True Origins of Man!

For the specific question that this Booklet examines – **Sir Isaac Newton's "Plan Of The World"** – the *knowledge* of Creation and Evolution which Fenton *illuminates* is key to *further* understanding that *that* knowledge *concomitantly* *gives* a clear *picture* of the humanly-unfathomable reality of the **'Structure of Creation'**.

The factual essence of that **"Structure"** is crucial to understanding why Newton sought so assiduously in **The Book of Revelation** for the **"Plan Of The World"**. The Bible is thus essential to understanding the definitive *cosmological connection and association* of Newton's **"Plan"** to **'The Revelation'**. In this present time of greater astronomical discoveries, we can now *know* what Newton, the *scientist's scientist*, sought to unravel; i.e., *what he searched for.*

Fenton's translation and insights thus offer far greater enlightenment overall than do other mainstream Bibles. In that regard, therefore, **Ferrar Fenton's Bible** inherently encompasses a greater degree of **Foundational Science** than do other Bibles.

Pope Benedict XVI's Call to Science

On September 13th, 2006, in Regensburg, Germany; Joseph Ratzinger – Pope Benedict XVI – in an address at the university where he was once a professor, established the basis for dialogue between cultures and religions; **in effect a new relationship between faith and reason.**

> "...only if reason and faith come together in a *new way*, if we overcome the *self-imposed* *limitation* of reason to the *empirically verifiable*, and if we once more disclose *its vast horizons.*"

More importantly in the context of the subject matter of this Booklet, however, the Pope pointed out that modern scientific reason:

> "...bears *within* itself a question which points *beyond* itself and *beyond the possibilities* of its *methodology.*"

> "Modern scientific reason quite simply has to accept the rational structure of matter and the correspondence between *our spirit* and the *prevailing rational structures* of nature as a *given*, on which its methodology *has to be based.* Yet the question *why* this has to be so, is a *real* question, and one which has to be remanded by the natural sciences *to other modes and planes of thought* – to *philosophy and theology.*"

"The West has long been endangered by this aversion to the questions which underlie its rationality, and can only suffer great harm thereby."

"The courage to engage the *whole breadth of reason*, and not *the denial* of its *grandeur* – this is the program with which a theology grounded in biblical faith enters into the debates of our time."

Comments of the Bishop of Rome

"In the Western world it is widely held that *only positivistic reason* and the forms of philosophy *based on it* are **universally valid.**"

"Yet the world's profoundly religious cultures see this *exclusion of the divine* [The Divine] from the *universality of reason* as an attack on their most profound convictions. A reason which is deaf to the divine [The Divine] and which relegates religion to the realm of subcultures is incapable of entering into the dialogue of cultures."

(All emphases mine.)

This address by the Pope and the concomitant statements by the Bishop of Rome are most timely given that certain earth-sciences are digging themselves deeper and deeper into paths that carry them further and further from the connection to natural truth set in place by the very processes of Creation itself. To that end, the subject matter of this Booklet perfectly resonates with the truths expressed by Pope Benedict XVI.

For, as we state unequivocally in **all** publications of **Crystal Publishing** – there should be no contradiction between science and Spiritual-Law truth, and that **science cannot supersede any such truths.**

8

The great scientist, Einstein, though noting the demarcation between science and religion, nonetheless understood the strong reciprocal relationships and dependencies between the two. He stated:

> "Though religion may be that which determines the goal, it has, nevertheless learned from science, in the broadest sense, what means will contribute to the attainment of the goals it has set up. *But science can only be created by those who are thoroughly imbued with the aspiration toward truth and understanding.*"

On pages 42-3 in *Ideas and Opinions*, Einstein summarises his view of the relationship between science and religion thus:

> *"Science without religion is lame, religion without science is blind."*

Introduction

The subject matter of this Booklet is very substantially derived from its Parent publication:

BIBLE "MYSTERIES" EXPLAINED
[Revised Second Edition]
Understanding "Global Societal Collapse" from
The "Science" in The Bible
What Every Scientist, Bible Scholar and
Ordinary Man Needs to Know!
Specifically from Chapter 11:
The "Seven Churches" In Asia: The "Revelation".

A stand-alone Booklet deriving from that Chapter – here re-titled **The "7 Churches" Of The "Revelation"** – might indicate that the subject matter within centres solely on religion. The Title of the Parent Work may also *seem* to suggest that. However, if we now add a sub-title which reads: — *What the "Hubble" Will Never See*: **Sir Isaac Newton's "Plan of The World"** — what might that addition *actually* suggest?

It strongly suggests that this Booklet takes you, the reader, on a journey that dares to push certain, current *scientific notions* about space to extraordinary and genuinely mindblowing conclusions. Moreover, whilst that journey *begins* with the scientific views, mathematics and theories of Astronomy/cosmology current in early 2009, it ends with the *astronomical* revelation that the great scientist, Isaac Newton, *intuitively* knew to be present in **The Book Of Revelation**.

For, as we have stated strongly and often in the Parent Work, The Bible needs to be recognised as **"A Primary Book of Foundational-science"** – *first and foremost*. All other considerations such as it being an historical and/or religious work, even though also true, are nonetheless subordinated by the clear explanations within it describing Creation, Creation-Law, evolution, anthropology, palaeontology, the Law of Numbers in mathematics, Plate Tectonics and — **COSMOLOGY**. Cosmology, however, on a scale not yet recognised by present-day astronomers. Intuitively recognised, however, by the 'scientist's scientist' — **Sir Isaac Newton**.

'Space! The Final Frontier!' The iconic expression associated with **Star Trek**, the space adventure series known to millions of fans world-wide. But is *Space* really the *Final Frontier* for we human beings of planet earth? The educational direction of present-day astronomers and their opinions sheltering under the broad-brush heading of 'cosmology' certainly suggests so. However, if we are open-minded enough to let go any and perhaps *all* pre-conceived notions about the *incomprehensible immensity* of 'Space! The Final Frontier!' – and, instead, embrace the very thing that the great scientist, mathematician, astronomer and theologian, **Sir Isaac Newton**, sought during his lifetime but did not find – we will begin to understand *why* he sought so assiduously for **"The Plan of the World"** in **The Book of Revelation**!

Noted for its *seemingly* indecipherable texts, an intimated connection between **The Bible** [specifically **The Book Of Revelation**] and earth-science cosmology will surely be construed as harbouring religious connotations. On the face of it, then, a strange "marriage".

Even though knowledge of 'The Plan' eluded him in his lifetime, Newton nonetheless believed the Greeks knew the secret of it, and that it could be found in **"The Revela-**

tion". In this Booklet we will examine Newton's sure belief and actually *reveal* **"The Plan"** that Newton sought therein. Grounded on *his* insightful recognition, "Newtonian physics" defines for science what we know today as a sure truth: *"That only a few natural laws apply to the whole universe."*

Since the cosmos is *our* sure reality and that we exist in it, *simple* logic must surely presuppose a 'coming-into-being' of that material immensity – from somewhere. Either that, or it has existed forever; for 'eternity'. That being an untenable notion, brutal logic must therefore decree that 'the Universe', as we *believe* we know it, emerged from 'something'.
The Parent Work examines this "emerging" from that "something" in Chapter 2: **"The Origins of Man: Genesis and Science Agree!"** [1]

Now, the very fact that we must marry earth-science and the **'science'** of **The Bible** to discover Sir Isaac Newton's **"Plan"** naturally presupposes that the mind-blowing mathematics of present-day cosmology are *absolutely necessary* for our complete elucidation here. And that is so, for certain major discoveries in astronomy etc., are crucial to our understanding of Newton's intuitive insight. So, what 'academic qualifications' do I bring to this radical analysis? Whilst I am neither a cosmologist nor a mathematician, I *am*, nonetheless, a writer who dares to question science/theology as it is 'packaged' to the world today.

At present, academic achievements coupled with intellectual sophistry are lauded virtually above all else. These two 'cornerstones' of world and 'worldly' education hold powerful sway in probably most Universities, and thus strongly drive the 'learning paradigm' in many countries. Anyone writing 'scientifically/theologically', as it were, is invariably expected to have at least some 'University letters' after their name. On

---

[1] Also published as a stand-alone Booklet of the same name. Available in the U.S. at www.crystalbooks.org

that basis with regard to this Booklet and the subject matter it examines, one *could* therefore *perhaps* dismiss it without any examination whatsoever for, *on the surface*, it evidences not even a hint of *academic* qualifications *for the author*.

However, in the many and *seemingly* irreconcilable points of disputation between science and religion – in *this* case on the subject matter herein; in *this* Booklet we provide a particular and **decisive** insight that cuts completely across long-held scientific notions on **'space'**. And we do so without the need for 'letters' and/or 'educational titles'.

Despite the very large numbers of well-educated and 'many-lettered' cosmology-authors who continually *theorise* on the size, extent and make-up of the cosmos, no one theory ever solidifies into a decisive insight where *all* in the Discipline can finally say: "We now have the answer/s."
Or, at the very least, '**the final foundational-insight** to build **constant-upon-constant**'.

Quite logically, such answers would be absolutely relevant for **both** science **and** religion. The idea that science and religion should be mutually exclusive and irreconcilable is, in the final analysis, a foolish one anyway. In the case of *this* vital question, **it is singularly so**. For the Truth of it offers the key knowledge of humankind's *actual place* in the overall and *far greater paradigm* of **The Creations**!

With regard to *my* 'qualifications', therefore: In Truth, I possess no such academic refinement from any worldly "Institution of Higher Learning". What I do have *in the first instance*, however, is simple 'common sense' coupled with a very logical-thinking mind. For the particular subject that this Booklet examines, the 'logical-thinking' aspect is perhaps brutally so. But therewith, precisely the *'right stuff'* to strip away the strong views of cosmological-science to *thereby reveal* what the great scientist, Sir Isaac Newton, sought: **"The Plan of The World"**.

In the pages of **this** Booklet, therefore; you, the reader, **will discover this clear fact for yourself.**

Given its bold standpoint and associated challenge to long-held and sacrosanct beliefs from main-stream "Christendom" about Creation and the material universes, and similar kinds of scientific mind-sets about the universe/s, the 'learned' from main-stream Christian/religious and scientific/cosmological academia, will almost certainly and predictably **wrongly** determine the analyses in this Booklet to be incorrect. Moreover, it will surely be labelled foolish by some, perhaps even by many.

So, *before* we *seriously begin* our contentious analyses: Do I possess *any* mandate at all? Yes, I most certainly do! But not from a University or similar. For this work, I accept the mandate given by Paul, the Apostle of **Jesus**.

Echoing Isaac Newton's mandate, I therefore remind all readers, especially cosmologists and theologians, of Paul's key statements to his followers. Accepted as a noted *intellectual thinker and scholar* by Christian academics throughout history and by many of the same today, let us all take note of Paul's clear admonition to the academic elite of *his* time and apply it to the *present*, and thus to the *author* of *this* Work."

Therefore: From Paul, founder of "The Church"; for University academics, for Christian theologians and for the reader of this Booklet; I accept as **my** clear mandate the directive **he** gave to **his** followers.

From 1 Corinthians, 1:26-29, in Ferrar Fenton's amazing Work: **"The Holy Bible in Modern English"**:

> "For, contemplate your vocation brothers: that not many philosophers, not many powerful, not many highborn – on the contrary, God has **chosen** the **foolish of the world**, so that He might **shame the philosophic...**"

The Jerusalem Bible states it similarly.

> "...how many of you were wise in the ordinary sense of the word, how many were influential people, or came

from noble families? No, it was to *shame the wise* that God *chose what is foolish by human reckoning...*"

(All emphases mine.)

Since I have no academic mandate either from a 'Theological College' or 'Scientific University', I will, for the subject matter herein – which greatly derives from The Bible and thus the *'science'* that is its *actual foundation* – therefore accept without boast the *greater mandate* from Paul, the appointed Apostle of that time.

And since Pope Benedict has made a logical plea to scientists that science and religion/spirituality should not be mutually exclusive, here we should therefore also include two very relevant texts for earth science: One from 'historical science'; the other from The Bible.

At present, the current "standard scientific viewpoint" fails to take into account or even acknowledge that all scientific endeavour must also embrace the reality of the non-material aspect of Creation that "**its concomitant Law**" unequivocally attests to. If it is ever to offer complete and meaningful answers to the never-ending questions it [science] continually finds itself faced with, it must find the courage to recognise the truth of what Dr. David Suzuki points out on page 19 in his book, 'The Sacred Balance':

> "Scientism, the aura of authority carried by scientists, has made us believe that knowledge obtained by scientists is the ultimate authority, that as we accumulate information, our capacity to understand, control and manage our surroundings will grow correspondingly. But the basic principle of scientific exploration contradicts this faith: knowledge comes from empirical observations, which are "made sense of" by hypotheses, which in turn can be experimentally tested. All information is open to being disproved. As Jonathan Marks has pointed out":
>
> "...the vast majority of ideas that most scientists have ever had have been wrong. They have been refuted;

they have been disposed of. Further, at any point in time, most ideas proposed by most scientists will ultimately be refuted and disposed of... Science, in other words, undermines scientism."

The message that the Great Prophet Isaiah gave millennia ago when he foresaw the effect that Plate Tectonics would have on the earth also thus clearly foresaw the quandary that would develop for earth-science and scientists if the higher knowledge of **Creation-Law** was not taken into account. Preceding Isaiah's visionary scientific-reality of a future "Reeling Earth" by thousands of years, the following ode could almost be called: **"The Scientists' Lament".**

Denunciation of Hypocrisy.

Now the Almighty demands,
"Why do this people approach
With their mouth and their lips,
To pay honour to Me,
While their heart is far off?
Their reverence is worthless to Me:-
It teaches the doctrines of men!

"So on this race I lay wonders,
Add wonders to wonders,
*Destroying its scientists' science,
And baffling its scholars' researches."*

(Isaiah. Ode 44: Book 1, Fenton.
Emphases mine.)

Chapter 1 of the Parent Work lists nine **"Crucial Imperatives"** that earth-science must engage with if that Discipline is ever to bring to an end its now ingrained ethos of believing that its constant intellectual theorising will somehow produce foundational answers that will not require further theories. So, in the case of the great cosmological question; finite or infinite?:

Crucial Imperative No 5:

That because the physical Universe is a *material* expanse, it is therefore *not* without end. **It is finite!**

In necessary concert with that reality, in association with **The Book of Revelation** to which we apply 'cosmological mathematics' to extract Sir Isaac Newton's **"Plan of The World"**:

Crucial Imperative No 1:

That "**The Bible**" should not be regarded as simply a religious work. The Bible should be *recognised* as a *scientific Work* for all of humanity, for it is a *Book of Spiritual and scientific Truth and Law*!

And also in necessary concert; this time, however, with Newton's stated recognition of immutable Law: That only *"a few natural laws apply to the whole universe"*.
He thus regarded those natural laws **"...as proof of the existence of a great and All-Mighty God"**.

In very necessary reinforcement for the subject matter herein, we repeat Newton's crucial recognition that:
'A rational God made a rational universe.'; and:
'All wisdom lay in the knowledge of numbers.'

[Whilst Newton clearly understood the wisdom inherent in the 'knowledge of numbers'; as applied to everyday earthbound 'mathematics', however, the higher reality of that 'knowledge' lies in **The Law of Numbers** both *in* and *as* its inherent and concomitant *Creative* aspect.]

Crucial Imperative No 6:

That there are certain and precise *Inviolable Laws* which govern *all life* and to which *all* human decisions and processes *are subject.* In their inherent Perfection these Laws are, in their perfect outworking, **Absolute**. And are therefore **Immutable**; and thus **Unchangeable**!

May the collective spirit of cosmological and religious academia be *well open* to the mind and spirit-extending ramifications of the following, *immutable*, "**Revelation**"!

The "7 Churches" of The "Revelation"...

"Happy are the reader and hearers of this prophecy who observe its records; for the time is at hand."

(Revelation 1:3, Fenton.)

"If I have seen further than other men, it is because I have stood on the shoulders of giants."

(Sir Isaac Newton.)

Crucial Imperative No 5:

That because the physical Universe is a material entity, it is therefore not without end. It is finite!

(Author. [Parent Work.])

In the Parent Work we examined The Laws of Life, our Origins, the processes of Death, the vital working of the Forces of Nature and noted the great Truth-Bringers; the *true* Teachings of whom we should have long-heeded. Since much of what we looked at was meant to be examined and learned in the *earthly* environment, this Booklet appropriately offers *ultimate insight* into the wider *material* environment to which the earth obviously belongs. It reveals the *limits* of the *material parameters* in which we live, *thereby* allowing us to know our respective place within those far-reaching, but nonetheless still just physical, bound-aries.

The truth that **Crucial Imperative No 5** inherently encompasses therefore sets precise boundaries on the otherwise humanly-incomprehensible immensity of the Material World.

"That because the physical Universe is a material entity it is, therefore, not without end. It is finite!"

The key focus of this Booklet is derived primarily from the revealing knowledge of **The Book of Revelation**, for *it reveals* the wider *material environment* that the physical universes *actually represent*. Precisely centred on the unfathomable vastness of that mind-numbing "amphitheatre"; as we reveal in a step-by-step process Sir Isaac Newton's "Plan of The World" we, the complete *human entity*, will therefore find *our relative, physical, contextual place* in the humanly-incomprehensible *material entity* of "space".

So key to knowing our *actual* place in the *greater* scheme of things centres on the question of who and what we actually are; i.e., the duality of man. Where do we, in our physical form, fit in all of this? Are we little more than flesh, bone and blood entities animated by a computer-like brain with a heart-pump holding it all together; nothing more than that...? If so, what, then, is the end outcome? Just a hole in the ground with all *animating life-force* extinguished?

If space-scientists, especially, believe that; *why bother* with astronomy and cosmology at all? What would be the point; to learn about an immensity that is the greatest marvel in the *Material World* and at the end of it simply be *no more*? Why bother with studying the incredible nature of the incomprehensible universe/s if that is the believed end-result? **Surely a pointless exercise in what would logically extrapolate to being an absolutely pointless career!**

No! Such a view is truly untenable. The huge industry now growing around the human genome and the overall genome mapping project is also ultimately just about *physical properties only*. The human-gene paradigm that science is all agog over now is therefore *not about* 'real life' at all. The human entity must understand this primary fact and yet deal with the greater and more crucial dimension of its true nature – **that of The Spiritual** – precisely as the present Pope has challenged science to open itself to.

Real life is therefore centred on our Spirit, the inner animating core within each of us, *upon which human genes have no bearing or influence whatsoever*. The opposite, however, is intrinsically correct. The spirit is the actual power and life-force within and thus ultimately drives *all movement and processes* concerning the physical body. The make-up and function of the very many genes that supposedly command all human life-processes, in the final analysis *all die with the physical body*.

National Geographic queried *exactly* this question in two separate Documentaries: "**Birth of Life**", and "**Human Ape**". The associated and most relevant point asked was:
"**How did *non-living* material come to life?**"

The History Channel, too, sought the same kind of definitive answer in the series, "**How Life Began**". It asked:
"**Where did [this] life come from? What IS life, exactly?**"

And in a *space* of perhaps *insightful prescience*, the Series further and crucially queries:
"Is it chemical, spiritual, or a combination of both?"
[All emphases mine.]

The same Documentary Channels also continually screen scientifically-updated programmes about 'space'. From various key universities, many experts in this field offer their particular level of 'expertise' to the documentaries – astronomers, cosmologists, astro-physicists, astro-biologists etc., etc.. What is most striking with regard to this group is the fact that a 'collective', across-the-board, academic notion has seemingly emerged from *their* study of the stars. Currently taught to 'space' students globally, the overall paradigm basically states:

1. That the 'creation' of the *physical* universe singularly provided, *on its own*, the building blocks for *all* life to emerge – including us.

2. That the increasing knowledge about the physical properties and make-up of the stars in the universe, including the various gases within it, therefore represents the greatest discovery of all for the human race.

3. And that, concomitant with such a view – now part of the educational-paradigm of earth-science – we, the 'complete' human entity, are thus constituted of star-material, solely. Stardust.

4. Within that strongly-promoted, so-called, "scientific reality", therefore; no place exists or can exist for an *inner animating core* which would also necessarily be the 'separable', *non-physical*, 'life-force' *within* and *for* the human entity.

For many astronomers and cosmologists it would seem; no belief in a **life-animating** 'soul' or 'spirit' most necessary for one's "ongoing existence". Just 'stardust'; i.e., atoms and molecules precisely configured – or perhaps *re-configured* – to form the flesh, blood, bones and organs necessary for physical earth-life, but which must then somehow *animate itself* so as to become a living, breathing, internally-pulsating, communicative, mobile human being. Shall we repeat the key question? We should!

"How did *non-living* material come to life?"

So on the basis of reconfigured 'stardust'; at the end of a lifetime of valuable furtherance of human knowledge, no 'form' that continues on. Just a black hole of nothingness where the very thought-processes that recently engaged with concepts which could not have been even imagined just generations ago are extinguished.

Gentlemen and women of space research: Sorry! Such a notion is not only untenable, it is really illogical. It is, in the final analysis, the ultimate *'nonsense-notion'*, for *inanimate matter* cannot possibly *animate **itself***, regardless of what empirical science might determine as 'its truth'. Moreover, that is a *sure reality* which you will *all* discover to be *true* upon *your* exit from earth-life.

The very fact that cosmologists and astronomers do not yet recognise the ***full extent*** of the ***material universes*** – and perhaps never will – means that *that* essential knowledge is therefore denied to students under their instruction. Yet such knowledge is imperative for the human entity to know of in order for *this* fact to "fully hit home":
That *as incomprehensibly vast as what we will reveal further on, the unfathomable scope and scale of it all is nevertheless **just our physical home for our physical form**.*

Therefore, notwithstanding its humanly-incomprehensible size, the total *expanse* in which this far greater reality exists ***is*** the ***smallest*** and ***lowest*** part of all Creation! [Newton's Law of Gravitation.][2] In *this* Booklet, from The Book of Revelation, we will elucidate exactly that crucial knowledge.

Only by that very sobering realisation can we gift to ourselves inner peace deriving from the fact that "**home**" truly is far *above* the Material Worlds. In Realms once known by earlier humankind to be sure fact – before being *intellectually relegated* to a 'dustbin of derision' by the blinkered, intellectual constraints of 'earthbound science'.

Astronomers and cosmologists, therefore: Do not place yourselves in the position of the Scribes and Pharisees of Je-

---

[2]Explained more comprehensively in the Parent Work: BIBLE "MYSTERIES" EXPLAINED ... Chapter 2: **'The Origins of Man; Genesis and Science Agree'**. [See end of Booklet], and in a standalone Booklet of the Chapter Title.

sus's time; they who sought religious/academic control over the masses. Remember His warning to *them*:

> "Woe to you, play-acting professors and Pharisees! because you lock up the Kingdom of Heaven in the face of mankind; while you yourselves neither enter, nor allow those arriving to go in."

(Matthew, 23:13. Fenton.)

Instead, open up to the all-encompassing **truth** of the **extent** of the physical Universes [plural] that your great "astronomy ancestor", Newton, *intuitively understood* to be factual reality. Though eluding him **then**; today, with the *immutable knowledge* of **Creation-Law** coupled with recent huge strides in cosmology, *you* can finally know: —

0.1 Isaac Newton's "Plan of The World"

The Book of Revelation, by virtue of its seemingly enigmatic content, provides fertile ground for many interpretations, from the literal to the bizarre. That particular book is perceived to cover many prophetic aspects, though often *apparently* unrelated, and *ostensibly* without clear linkages. 'The Revelation', being part of that Work which is the foundation of Christianity – the be-all and end-all by their own admission for this particular group of global humanity – is nevertheless considered by many to be virtually impossible to understand logically. Yet if the purpose of The Revelation **really was** to help **all** of humankind **understand** our 'reason for being' at precisely **this present time of our tenure in the Material World** what, then, must derive from either a wrong interpretation, or from ignoring it completely?

As we stated in the Parent Work and repeat here, during his time with the Church the great mathematician, Isaac Newton, directed his monumental talent of genius to analysing The Bible, trying to discover the secret knowledge he believed lay hidden there. He further believed that some of the ancients – in particular the Greek mathematicians – had known this secret. If he could find it, he would know it too:

The Plan of the World!

As *pointedly* stated in the **Introduction**, the recognition of two key factors lay at the heart of Newton's unshakeable belief:

1. A rational God made a rational universe; and

2. All wisdom lay in the knowledge of numbers.[3]

Of special interest to Newton was The Book of Daniel with its mathematical time-line of prophecy. The Book of Revelation, however, was the *primary* Book of The Bible wherein Newton sought his 'Plan'.

The journey we need to take in order to clarify the "7 Churches" *mystery* has powerful echoes to Newton and his genius, for he provides the foundational mathematics of astronomy which we will need to arrive at our conclusion. Why astronomy for that conclusion? For the moment we should let that reason be a 'revelation' in itself. Since we have Newton as a man of the Church and a towering genius of science — surely the quintessential *"scientist's scientist"* — he is the ideal companion for this august journey. Let us, then, not just take this man of spiritual intuition and intellectual genius with us, but let us stand on *his* shoulders and discover the treasured goal that he could not find in his lifetime: **"The Plan of the World"**.

Since Newton believed that *"only a few natural laws apply to the whole universe"*, he therefore regarded those natural laws **"...as proof of the existence of a great**

---

[3]The term "knowledge of numbers" in the earthly *mathematical* sense, is, in the higher *degree* of knowledge, actually **"The Law of Numbers"**.

27

and All-Mighty God". Precisely *because* he was both a theologian and scientist vitally interested in apparently all things, including The Revelation, we would sincerely hope that modern-day scientists and theologians would want to journey with us to also discover **'The Plan'**.

Any notion that states The Book of Revelation as being an enigma and therefore indecipherable, must thus presuppose that some of the greatest answers to life which lie in there might be too difficult to even attempt to understand. In our view that is an illogical and untenable position. After all, it is "The Revelation". By such statements, we reveal the actual level of **our** *non-understanding.*

As has been pointed out in the Parent Work, the very fact that we are here on planet earth must logically infer that we are **meant** to discover **all** the answers to our purpose for being. Otherwise it all becomes a rather pointless exercise: of either scientific contention deriving mainly from a theoretical-supposition basis, or religious contention deriving from differing interpretations of the many and varied beliefs awash across the globe. Unfortunately, ideas, theories, guesswork etc., do not provide *definitive* answers, a good reason why the admonition, "Seek and you shall find", is so vital for we of the human race. This most necessary *'finding'* of the final and ultimate answers to the questions to life cannot be achieved with the use of our intellect alone, however.

For the complete understanding of this particular Booklet, we must *strongly reiterate* that the purpose of the intellect is to facilitate, to the highest possible level, the material and technological undertakings that humans require for their sojourn and ongoing development in the *earthly environment.* The intellect possesses no understanding of the many and far higher non-physical Realms of the incomprehensible total we sometimes too loosely call Creation. That is because *only* the *physical-body* part of man is *derived* from the *material* world.

So also, therefore, is its closely associated aspects of intellectual brain-activity. Thus the 'seeking' of final and complete knowledge must be driven by the *Spirit* – because it is connected to The Source of Life. Such seeking, however, needs a *clarified* intellect as its companion. In any case, no amount of intellectual or theological sophistry can change *what actually is*. So any religious or scientific pre-conceptions brought to bear here to attempt to discredit the true meaning of the "7 Churches" question is, in the singular nature of things pertaining to **Creation-Law Truth**, *immediately rendered irrelevant*.

Anyhow, it is well past the time that the true meaning of the "7 Churches puzzle" was recognised. That essential recognition, as a vital part of the complete knowledge for humankind, permits critical threads to many of the key questions of life to be woven together to reveal a far larger picture than could ever be the case without this information. Moreover, it also provides, perhaps paradoxically, the right 'structure' – from the *earthly* viewpoint in this case – for the very necessary, and greater, understanding of a singularly-defining event in the history of mankind: The true meaning and purpose of: **The earthly Birth and Life of Jesus!**

In the Parent Work we explained the process of the formation of the various levels of Creation in broad outline.[4] Now we need to do the same here, but solely with the *physical universes* of the *Material* part of Creation. In this case by using the mathematical unit of the 'light-year' as our measuring staff.

0.1.1   The 'Mathematics' of Cosmology: The "Big-Bang" and "Inflation" Model

So, to arrive at the point of sure and conscious knowing about the meaning of the "7 Churches", we need to now extrapolate our present view of the physical universe/s upward and outward. Firstly from the immediate environs of our solar system, and thenceforth undertake a mathematical journey

---

[4]See Table of Contents of Parent Work at end of Booklet. Specifically Chapter 2. Also a stand-alone Booklet.

into distances so vast as to be totally *incomprehensible*. Even the word itself does not nearly suffice to describe the immensity of just our 'home galaxy', let alone the stupendous nature of what is *observable* beyond that; what science *believes* is the *complete* universe.

On the question of the very much vaunted 'Laws of Physics' – which can *supposedly* explain *all* events and processes – we should note that the current 'scientific mindset' centred on the *validity* of those laws, particulary in cosmology, *fail completely* when confronted by the 'perfectly natural' cosmological *reality-paradigms* of quasars and pulsars. And certainly in the case of the dreaded 'black holes' now known to exist at the center of most galaxies.

Now why should that be? Why should the mathematics of cosmology, which has answered many 'space questions' thus far, now fail in the face of those truly amazing spectacles? Has earth-science reached a point where it is simply 'outgunned' by cosmological processes far beyond any human empirical understanding at this time because its *current level* of mathematics is completely inadequate for such an *especial degree* of elucidation of what must nonetheless be *perfectly natural processes* – albeit, however, on a truly gigantic scale? For astrophysicists, therefore, could it be nothing more than just **"The Error of Scientism"**, which we quote here? From page 19 of 'The Sacred Balance' by David Suzuki:

> "Scientism, the aura of authority carried by scientists, has made us believe that knowledge obtained by scientists is the ultimate authority, that as we accumulate information, our capacity to understand, control and manage our surroundings will grow correspondingly. But the basic principle of scientific exploration contradicts this faith: knowledge comes from empirical observations, which are "made sense of" by hypotheses, which in turn can be experimentally tested. All information is open to being disproved. As Jonathan Marks has pointed out":

"...the vast majority of ideas that most scientists have ever had have been wrong. They have been refuted; they have been disposed of. Further, at any point in time, most ideas proposed by most scientists will ultimately be refuted and disposed of... Science, in other words, undermines scientism."

As we must constantly reinforce, **The Spiritual Laws of Creation**, alone, hold the final keys to understanding the true nature of 'The Universe/s'. Earth-based university-mathematics, whilst necessary for most 'material' applications, are a very poor cousin to **The Law of Numbers**. It is thus the far *Higher Laws* which *entirely govern* all events and processes in "**The Creations**".

To that end, let us examine the 'mathematics of cosmology' to begin with, and then travel far further with **The Law of Numbers** to learn the **"The Plan of The World"**; the 'secret' which Newton sought in the pages of that premier foundational-book of science: – **The Bible**. So:

One light-year represents the distance that light covers travelling in a vacuum for a period of one year – approximately 9.4607×10^{12} kilometres $(5.878 \times 10^{12}$ miles, at a speed of 186.000 miles per second).

Our journey starts, naturally, from Earth. It is the third planet revolving around a relatively small sun in a solar system residing at the outer edges of a galaxy designated the Milky Way Galaxy. It contains 100,000 million odd stars to which the sun of our Solar system belongs. Even travelling at light-speed, sunlight takes 8 minutes to travel the 150,000,000 km before reaching us. The same light travels 5 more hours before striking the planet Pluto at the outermost edges of our Solar System. And 4.3 light-years later, or 40 trillion kilometres away, it reaches our nearest stellar neighbour, Alpha Centauri.

Our galaxy, a disc-shaped collection of stars with the Earth about a third of the way out from the centre, was once thought to be the entire universe until discoveries in the 1920s revealed a far greater expanse beyond it. Today we know it is only one of billions of galaxies. An observer looking at the Milky Way from earth is actually looking edge-on into the Galaxy.

In a broad-brush time-sweep, the Galaxy started to form some 10,000 to 14,000 million years ago, and its oldest stars are estimated to be perhaps up to 15 billion years old. Our sun takes about 230 million years to complete one journey round the centre. Our galaxy is about *100,000 light-years in diameter*.

Beyond the 'Milky Way' can be located galaxies in every direction. We are part of a loosely bound cluster of some 20 galaxies called 'the local group'. From the centre to its outer boundaries is roughly 2,000,000 light-years – *4 million light-years across*.

The next larger formation we belong to is known as a 'local supercluster'. Clusters of galaxies – like fleets of ships – congregate in superclusters. The closest cluster to our local group is some 50 million light-years away, near the centre of our local supercluster. From the centre to its outer boundaries is roughly 75,000,000 light-years – *150 million light-years across*.

Deriving from the present level of earthly 'cosmological knowledge', the only step left to take now with the present level of earthly knowledge derived from astronomy and cosmology, is into what is termed 'the known universe' – the largest expanse by far. It is that of the farthest reaches of the universe which can be observed by the use of optical or radio telescopes. Our universe is stated to be isotropic in nature and form which means it looks the same in every direction. Quasars are the most distant objects observed. Each of the brightest quasars emits the energy of hundreds of galaxies from a volume far smaller than our Milky Way Galaxy. The furthest quasars are stated to be rushing away from us at 90% the speed of light.

The 'known universe' was, until quite recently, believed to be about *40,000,000,000 light-years across*. Relatively recent estimates to the *edge* of the universe – at least the visible part of it – determined it to be **100 billion trillion kilometres away**. Further recent estimates, however, now calculate the *size* of our universe at roughly **100 billion light-years across**.

So how did that vast, utterly incomprehensible expanse come into being?

The question of how the universe came into existence only gained real traction from the early 20th century. Notwithstanding Newton's crucial contribution to the science of astronomy, still relevant today – NASA acknowledges Newtonian physics as the foundation for its space programme – it was not until the larger telescopes and more powerful computers were developed that astronomers began to get a sense of how it *might* have begun and how it all *might* work.

At this present time the 'Big-Bang theory' holds sway. Other ideas have been mooted and subsequently discarded. Various possibilities have encompassed an 'ever-expanding universe', the 'steady state' theory and an 'open universe' – whatever those terms *really* mean. Currently thought to have brought the universe into existence, the rather impossible-to-grasp sums of the "Big-Bang" explaining how it all began and developed, at the very least make fascinating reading.

On the question of the merits or otherwise of the 'Big-Bang', the reader must obviously decide for himself. Current *scientific* thought on the how and why of the universe, even though able to now answer many previously unanswerable questions, is nevertheless still *solely-derived* from just an *empirical* paradigm.

As we have previously stated, because the present level of 'scientific knowledge' is not based on a **Creation-Law** foundation to begin with, it does not recognise – let alone even begin to take in – the far *greater* expanse of the *non-material* worlds. Science is therefore *unable* to derive the final answers about the *physical universes* for, with its *present* level of knowledge, it can only engage with the 'singular universe' which the "Hubble" has mapped. Not possessing the requisite knowledge and thus recognition of the far greater and higher paradigm encompassing *all* the *non-material Realms* upwards to our Spiritual Origins, means that cosmological science has effectively shackled itself to a very constrained, solely-empirical, physical 'box'.

> We sincerely hope that at least *one* cosmologist *somewhere* will intuitively perceive the *Truth* of Newton's very sure recognition, and thus smash through the walls of *modern-day* cosmology's self-constructed and very myopic 'box'.

For only from out of that far greater paradigm – which encompasses all of the non-material worlds upward to humankind's true Origins *to begin with* – will cosmology ever be able to *completely contextualise*, and thus *fully answer*, the enigmatic 'riddle of the universe': Sir Isaac Newton's **"Plan Of The World"**.

So, the basic mathematics of the Big-Bang theory, according to more recent analyses, proposes something called "The theory of inflation" to explain its origin. From the feature article in Time, June 2001 ["How the Universe Will End"], it states that:

> "...the entire visible universe grew from a speck far smaller than a proton, to a nugget the size of a grapefruit, almost instantaneously, when the whole thing was 0.0000000000000000000000000000000001 second old." [Have I even quoted the correct number of zeroes here?]

According to the "inflation model" the incomprehensible immensity of the universe came from virtually nowhere in an instant. So small that you would have needed a microscope

34

to find it. The inflation theory was proposed in 1979 by Alan Guth, then a junior particle physicist at Stanford University. It holds that:

"...within a fraction of a moment after the dawn of creation[5] the universe underwent a sudden dramatic expansion. It inflated."

(Time, June 2001. Feature article...)

The whole episode evidently lasted no more than one million, million, million, million, millionths of a second – but it transformed the universe from something that could be held to something at least 10,000,000,000,000,000,000,000,-000 times larger. According to Guth's theory, gravity came into being at one-ten millionth of a trillionth of a trillionth of a trillionth of a second.

In a single moment, we were endowed with a universe that was at least 100 billion light years across. Feasible? Sounds impossible. For how can one individual second be logically and understandably carved up into such infinitesimal part-seconds to begin with; to explain something so explosively-expansive in the time-frame proposed? Anyway, who on this earth can finally say?

It is interesting to note the opinions of other astronomers, such as Martin Rees:

'Martin Rees, Britain's Astronomer Royal, believes that there are many universes, possibly an infinite number, each with different attributes, in different combinations, and that we simply live in one that combines things in a way that allows us to exist... Rees maintains that six numbers in particular govern our universe, and that if any of these values were changed,

---

[5]The 'Creation' that astronomical science refers to here represents just the *material worlds only* – the *smallest* and *lowest* part of **Creation** proper; Newton's Universal Law of Gravitation.

even very slightly, things could not be as they are. For
example, for the universe to exist as it does requires
that hydrogen be converted to helium in a precise but
comparatively stately manner – specifically in a way
that converts seven one-thousandths of its mass to energy.'

Lower that value very slightly – from 0.007 per cent
[sic, should be 0.7] to 0.006 per cent [sic, should be
0.6] say – and no transformation could take place; the
universe would consist of hydrogen and nothing else.
Raise the value very slightly – to 0.008 per cent [sic,
should be 0.8] – and bonding would be so wildly pro-
lific that the hydrogen would long since have been ex-
hausted. With the slightest tweaking of the numbers
in either case, the universe as we know and need it
would not be here.'
[Figures clarified from "Seven Wonders of the Cosmos",
p.200 (See Biblio.)]

(How to Make a Universe. Reader's Digest.
August, 2004)

What we thus clearly note here is a *perfected state* for life
to be able to exist at all, and a stupendously huge "home",
in the physical sense, for it to exist in. The Perfection of Creation overall surely precludes any notion that it all emerged
by "accident".

The question of what might happen if we were *able* to
travel to the edge of the universe and look beyond it is an
interesting though perhaps pointless 'scientific quandary' ultimately. Nonetheless, Einstein's Theory of Relativity holds
that the universe bends in a way that cannot adequately be
imagined so we would, even after travelling in a straight line
to the edge, eventually arrive back at our starting point. His
theory suggests that space curves in a way that allows it to
be boundless – **but *finite*.** So the physical universe we see
and *believe* we know **is finite!**

'Physicist and Nobel laureate Steven Weinberg explains
that space cannot even properly be said to be expanding, because "solar systems and galaxies are not

expanding." Rather, the galaxies are rushing apart. It is all something of a *challenge to intuition*. For us the universe goes only as far as light has travelled in the billions of years since the universe has formed. This visible universe – the universe we know and can talk about – is a million million million million (that's 1000,000,000,000,000,000,000,000) kilometres across. According to most theories, however, the universe at large – the meta-universe, as it is sometimes called – is vastly roomier still. According to Rees, the number of light years to the edge of this larger, unseen universe would be written not "...with ten zeroes, not even with a hundred, but with millions". In short, there's more space than you can imagine already without going to the trouble of trying to envision some additional beyond."

(How to Make a Universe.
Italics mine.)

To achieve the incredible expansion proposed by the "theory of inflation" – from an invisible speck to a structure billions of miles across in a fraction of an instant – the speed of light is very obviously totally inadequate. Einstein's Theory of Relativity, however, seemingly permits the mathematics to fit a "faster-than-light" possibility.

"An equally unsettling implication is that the universe is pervaded with a strange sort of "antigravity", a concept originally proposed, and later abandoned, by Einstein as the greatest blunder of his life."[6]

This force, which has lately been dubbed "dark energy", isn't just keeping the expansion from slowing down, it's making the universe fly apart faster and faster all the time, like a rocket ship with the throttle wide open. It gets stranger still. Not only does "dark energy" swamp ordinary gravity but an invisible substance known to scientists as "dark matter" also seems

---

[6] Adam Riess, a Space Telescope Science Institute astronomer, has seemingly helped prove that Einstein may have been right in the first place; a mysterious antigravity force that acts like Einstein's cosmological constant is evidently quite real.

to outweigh the ordinary stuff of stars, planets and people by a factor of 10 to 1. "Not only are we not at the centre of the Universe," University of California, Santa Cruz, astrophysical theorist Joel Primack has commented, "...we aren't even made of the same stuff the universe is."

These discoveries raise more questions than they answer. For example, just because scientists know dark matter is there doesn't mean they understand what it really is. Same goes for dark energy. "If you thought the universe was hard to comprehend before," says University of Chicago astrophysicist Michael Turner, "...then you'd better take some smart pills, because it's only going to get worse."

(Time, June 2001. Feature article –
"How the Universe Will End.")

From two fairly recent publications, we have noted and quoted the opinions and theories of some key academics in the field of cosmology and astronomy etc.. So it seems the "Big Bang" is the theory that is current, and ongoing research appears to support and strengthen it. Thus, from an infinitesimal speck, the huge and completely incomprehensible size and mass of the universe was, by cosmologists' reckoning, "suddenly there".

The key question that obviously arises is: How could this incredible mass emerge from virtually nothing? Even if we use a standard analogy of growth – say that of a human being where, at conception, the potential is simply that of a very tiny, fertilised egg but where the full potential unfolds to adulthood – it is on a scale that is easy to understand. What about a very tiny seedling which might grow to become an immense tree? That, too, is easy to comprehend. Not the 'Big Bang' scenario, however.

The physical form of the human being eventually dies, decays and reverts to its component parts; earth to earth, dust to dust – as will the giant tree. And the mass of the earth, even with billions of creatures living and dying over aeons, remains the same. What about the total mass of *billions* of galaxies, each with their *billions* of suns, however? Can we really believe that *that* absolutely incomprehensible expanse could somehow explode out of something far smaller than a single dot on this page in the micro-millisecond time-frame proposed? Then very much later, in an equally incomprehensible future point in time, perhaps contract or implode, and squeeze itself back into something invisible, except through a microscope? It all seems too impossible, too strange, to *be* possible.

A 'white dwarf' is surely a good example; a very small, extremely dense star where the atoms in it have been broken up and the various parts packed tightly together with almost no waste space so that the density rises to millions of times that of water. According to scientific calculations, a spoonful of white dwarf material would weigh many tons. Neutron stars, made up principally or completely of neutrons, have even *greater* density. Clearly, individual stars can be compressed to an incredibly small size. And in a black hole where not even light can escape, even more so. Multiply the total mass of literally *billions of* **galaxies** *worth*, however, and how do they fit back into that point of almost nothing?

Irrespective of what may seem to be theoretically correct from the point of view of earth-science, every event that occurs can only take place within strict and absolutely lawful parameters. Therefore, is the 'Big Bang' theory correct? Deriving from theoretical analyses, empirical observations and mathematical calculations, is science right with regard to how it all came to be? Or has something more been missed or not understood?

0.1.2 The "Big-Bang": A Problematic Theory

The Parent Work [**Bible "Mysteries" Explained**] unequivocally states **The Bible** to be a *primary Work* of **Foundational Science**. In that key capacity, it describes concepts about, and concomitantly offers explanations for, the key questions to life. These crucial insights are vastly different to the standard educational fare fed to the student body of humankind. And in this Booklet you, the reader, will have certainly noted the clear fact that this writer finds the present state of general cosmological theory centred on the Big Bang illogical, and therefore untenable.

For cosmologists and astronomers, mathematics must fit any proposed theory, otherwise the model must be 'tossed out'. Given the scale and time-frame of the life of the universe/s – unlike the mathematics required for architecture and engineering – 'cosmological maths' must enter a very *different* realm. The necessary expansion of theoretical analyses certainly allow for radical proposals here, but the associated mathematics must somewhere, somehow, have it all 'make sense' in order 'to make it fit'.

Again, unlike architecture, which – if the maths are wrong might see the building collapse – the 'mathematical playground' of the cosmos will not produce 'collapsed buildings' for astronomers. Nonetheless, if something is not correct, it is *forever* incorrect; thus wrong!

As we must reinforce often, as long as astronomers remain locked to the notion that the be-all and end-all of human endeavour and existence is tied solely to just a Material-world paradigm, **they will never ever arrive at the correct picture.**

And therefore never recognise what the great scientist Sir Isaac Newton *intuitively* understood:
That the universal reality of **"The Law of Numbers"** inherent in **"The Book of Revelation"** can – if present-day astronomers/cosmologists were *also* intuitively *open* to such recognitions – actually reveal for them and the earth-science of astronomy **The Truth** of **"The Plan Of The World!"**

In April of 2009, the Telegraph Group news agency featured a spectacular picture of the collapse of a huge star.

40

The most distant object ever seen in the universe at 13 billion light years, it is so far away that its 'gamma ray burst of light' has taken almost the entire age of the universe to reach us. [Gamma ray bursts are the most luminous explosions in the universe and are the afterglow of dying stars.] Scientists believe the 'burst' was caused by a *massive* star collapsing and exploding at the end of its life, leaving a black hole. Yet we are told that the star which triggered this event, designated GRB 090423, was only 640 million years old.

Was this massive star 'unstable', and thus 'died young'? Or did it live out what would be a normal life span for such a large mass? Professor Edo Berger, from the Havard Smithsonian Centre for Astrophysics in Cambridge, Massachusetts, who also studied the burst, said:

> "We now have the first direct proof that the young universe was teeming with exploding stars and newly born black holes only a few hundred million years after the Big Bang."

In the ordinary course of processes cosmic we are told that a *small* star will live longer than a *massive* star. Simple logic would *seem* to suggest that the opposite would be the case. However, a larger body will burn off its *fuel supply* at a faster rate, thus leading to a 'short' life. Our sun has not only lived for 4.5 *billion* years and counting versus just 640 *million* in this example, but will be around for a while yet.

In that single fact lies the *key insight* into not only *why* our sun is the size that it is, but *therefore why* we as a human race *concomitantly inherited* such a vast time-span for life *down here*. The long evolutionary processes we needed to undergo required a time-period commensurate with not just our *physical* development, but more importantly our *spiritual maturation*. Therefore, only a small sun could provide the time required.

So, in the case of the earth, was it all a cosmic *accident* formed at random by countless trillions of cubic miles of 'stardust' somehow coming together in precise proportions to begin the thermonuclear processes which gave birth to our galaxy, our sun, our solar system; then earth and the millions of diverse *living* species on it? Or did it come into being through **Divine Ordination** gifting we humans *conscious* life? Thus: **A Plan!** Deriving from the logical analyses throughout the Parent Work, particularly, our unequivocal conviction embraces the latter view.

On the question of a problematic 'Big Bang'; according to **Daniel Pendick**, Associate Editor of **Astronomy Magazine**, much evidence exists to support the 'standard model', such as the 'radiation afterglow' from the Big Bang, known as the cosmic microwave background [CMB]. Maps of the CMB from the Cosmic Background Explorer [COBE] and, more recently, the Wilkinson Microwave Anisotropy Probe [WMAP] confirm a number of Big Bang cosmology predictions. However, trouble spots, such as the true nature of dark matter – the glue that binds galaxies together – remain. Strange patterns in the CMB challenge one of the foundations of the standard model, inflation theory. The 'standard model', moreover, requires innovative hypotheses 'to make it all fit'.

One of the strangest proposed is 'dark energy'. The notion of dark energy offered a ready explanation for accelerating cosmic expansion, yet 'creates nearly as many problems as it solves'. From page 48 of Astronomy Magazine, April 09, ref: "Is the Big Bang in trouble?"; Lawrence Krauss, a theoretical physicist and cosmologist at Arizona State University, says:

> "When it comes to dark energy, we know that it exists, but we don't know anything about it."

Whilst WMAP's detailed picture of the CMB, the supernova observations, and 'various surveys of the distant universe have advanced cosmology at light speed', those discoveries did not get to the heart of the matter.

> "We've been so successful that the questions we're ask-
> ing are *so deep* that they may remain *unanswerable*
> for some time to come – and *maybe forever.* We don't
> understand the model we have. It's completely inex-
> plicable."

(Krauss, p.48. Italics mine.)

The concept of 'inflation' resolved fundamental theoretical problems for Big Bang cosmology. Blossoming into 'multiple versions', which according to science rests on a *'solid*
foundation of *theoretical* physics' [surely there sits an amazing 'contradiction in terms'], there nonetheless remains an
unresolved question. That is the question that this writer –
who is neither an astronomer nor a scientist, and who has no
'letters' after his name – also states cosmology must unequivocally answer: "How do we know inflation really happened?"
Gary Hinshaw, an astrophysicist at NASA's Goddard Space
Flight Center in Greenbelt, Maryland, and a member of the
team that designed WMAP, on page 49 of that analysis says:

> "The idea that there was a period of exponential growth
> is by far the best explanation we have of the current
> data. ... But the details of inflation, we have very
> little grasp of."

Theoretically, inflation generated ripples in space-time
called 'gravitational waves', which would have left 'an imprint on the CMB'. According to Krauss [p.49, all emphases
mine], even if an 'imprint' were found, that might not be
sufficient to 'dispel all doubt'.

> "The issue is not whether it is consistent with obser-
> vations. ... The question is how to falsify it. What
> could you observe that would be *different* if inflation
> *did not happen*? It is a beautiful, natural explanation
> of everything we see, **but that doesn't mean it's
> right**."

The discovery of dark matter helps cosmologists understand how galaxies and galaxy clusters hold together. In

fact, dark matter's gravitational influence keeps them from flying apart. However, whilst it fits the standard model, no one has yet found a particle of the stuff. An alternative to dark matter, called 'modified gravity theory', holds that gravity behaves differently 'out there among the galaxies'. Even though this theory accounts for some of 'the same observations as dark matter', it nonetheless faces a 'major hurdle': Einstein's general theory of relativity. Einstein's theory explains the universe more comprehensively than 'modified gravity' does.

'Astronomy' reports that data culled from the CMB support key aspects of the standard model, in particular a snapshot of the seeds of cosmic structure that evolved after inflation. British theoretical physicist, Stephen Hawking, on page 49 'proclaimed the COBE satellite's first glimpse of the CMB' as, [very *materially* and thus *very wrongly* in our *seriously unequivocal* view]:

"...the discovery of the century, if not *all time*."

'But nothing is beyond question in the standard model. Since WMAP released its first set of data in 2003, scientists have found patterns in the CMB that seem at odds with the standard model'. 'If someone finds a one-in-a-million weirdness in the CMB – highly unlikely to be an accident – it might suggest something is wrong with inflation theory'. Hinshaw, page 50, says:

"The consequences are *potentially profound*. The large-scale alignments in the CMB might be *inconsistent* with *inflation*."

So: What fills the 'empty' parts of space, the incomprehensible areas in-between the equally incomprehensible galaxies?
The standard model hypothesises 'dark energy' as the intrinsic "vacuum energy" of empty space. This notion postulates that as space expands, the density of dark energy [and its

consequential 'repulsive effect'] remains constant, while matter [and its gravitational pull] thins out. Cosmic expansion thereby speeds up. General relativity and quantum mechanics offer a mechanism for calculating the energy of empty space. [Here we have one more example of impossible-to-understand numbers.]

The answer turns out to be 10^{120} more energy than astronomers have actually measured. That's the number 1 followed by 120 zeroes. Lawrence Krauss dismisses this quantity as "ridiculous".

At this point in Astronomy Magazine's article on the problematic "Big Bang" theory, further quotes from various theorists have resonance with the very thing we state must be taken into serious consideration by *all* scientists if they are ever to even get close to *really understanding* the 'space-time continuum' of The World of Matter. For that is our home 'down here', and in which we *must* reside for *part* of our *complete* existence.

To illustrate *our* point, 'Astronomy' notes that Krauss and other theorists find it *unsettling* that:

> "...the universe contains just enough dark energy to have allowed galaxies and other structures to form – and, coincidentally, human observers to exist."

Now why should the interesting term, *unsettling*, be used to describe what really should be easily recognised by *all* astronomers, cosmologists, astro-physicists – and whatever other names/titles this particular branch of 'human learning' deems relevant for its work – as the "created reality"?

In terms of the notion that dark energy actually exists, here we have only two possibilities:

- 'If there were a *small* amount of dark energy, or none at all; the universe would have collapsed in a *Big Crunch* early in the expansion.'

- 'However, if dark energy matched the incredibly large proportion predicted by quantum physics, it would have expanded so rapidly that nothing more than a thin fog of matter and energy would fill the universe today.'

Quite obviously, we live in a universe exactly tailored and proportioned to support the myriad life-forms present on earth. Here Astronomy uses another interesting term, *cosmic coincidence*, to explain why "...we seem to live in the best of all possible universes". The 'anthropic principle', invoked by some scientists to explain the 'odds stacked in our favour', states that '...the universe has the ideal amount of dark energy because we wouldn't be here to measure it if it didn't'.

> "If the amount of dark energy weren't that much bigger than what we measure, then there wouldn't be galaxies. And if there weren't galaxies, there wouldn't be stars; and if there weren't stars, there wouldn't be planets; and if there weren't planets, *there wouldn't be astronomers.*"

> (Krauss, p.51. [All italics mine.])

Daniel Pendick notes:
'The anthropic principle's explanation for dark energy has the potential to shake the foundation of all physics.' Page 51 of that analysis has Krauss stating: [Emphasis mine.]

> "Physics is supposed to predict why things are and why they have to be that way. This would say they don't have to be that way at all. **They just happen to be that way because <u>we're here</u>.**"

Whether from a foundation of physics, 'cosmic mathematics', or from simple but ultimately more valuable "intuition" driven by "the spirit" within each of us; in that last sentence Lawrence Krauss has hit upon the *true reason* for the existence of the cosmos in the first place.

We are not here by 'accident'. That is a truly ridiculous notion and a ***monumental error***. That idea, unfortunately

46

too prevalent in 'earth-science', concentrates *valuable* academic thought and research on a *completely wrong paradigm* which can *never* find resolution – for it is *forever* wrong.

Chapter 2 of the Parent Work: **The Origins of Man: Genesis and Science Agree**; explains why we are here and thus why we – the human entity – needed a material home for our physical component. Primarily, two words – or perhaps one hyphenated word – tells us why Lawrence Krauss is *correct*: **Free Will**; **Free-will!** Of all creatures in Creation, we, alone, possess the attribute of **Free Will!** That is why we seek answers to our reason for being. That is why we were seriously enjoined to:
"Seek and you Shall find."

Free Will, however, means exactly that! We are free to *choose* a correct path of learning – one that *will bring* the *true* answers and thus *genuine enlightenment*; or we can *choose* to follow a path that is literally ***a dead-end in all respects***. Many scientific paths are *already* dead-ends *even before* they start. In that regard, it is crucial for humankind to *get right* astronomy/cosmology, but in an *especial way* for a *most* especial *reason*.

At this time, that particular branch of earth-science correctly reveals the incomprehensible immensity of *our universe* in its *materiality*. However, what it does not do – and what really should be cosmology's next step and thus more *crucial* elucidatory purpose – is to recognise the connecting links that would offer exactly that especial 'cosmic revelation' we allude to.

For that to happen, however, *a fundamental shift in thinking must first take place within that scientific Discipline*. A paradigm shift – just to begin with – which encompasses a *greater* horizon than simply a *material* universe that astronomy *thinks* it knows. A *conscious* shift *into* the relevant *cosmological Creation-knowledge* would *bring* the requisite *recognition* that would allow the science to build *constant upon constant*. Thus *true knowledge* in place of the present, fractured state of that regime's 'guess-work reality'; trying to make things fit a particular theory or idea.

'Astronomy' gives an excellent example of differing ideas between the 'experts'. We quote the relevant segment verbatim:

0.1.3 Lost in the "Hubble Bubble"

'Some researchers have proposed to solve the problem by getting rid of dark energy entirely. It's theoretically possible to modify the standard model so that dark energy is not necessary to explain accelerated cosmic expansion. For example, Oxford University theoretical physicist Subir Sarkar and other researchers are investigating an alternative – the "Hubble Bubble" hypothesis. It holds that the local universe lies in a region of space with less-than-average density. It would expand at a faster-than-average rate relative to the space outside the bubble. If this is true, then accelerated cosmic expansion may be just a mirage caused by the assumption that the universe is homogeneous and isotropic on all scales. Critics say the Hubble Bubble hypothesis is an example of stacking up "what ifs" until they add up to the desired answer.' [Hinshaw says:]

> "You can contrive models to fit the data without dark energy, ... but it then becomes a question of what is really plausible."

'Dark energy, he says, is the most plausible of all known possibilities. But Sarkar insists it's too soon to dismiss alternatives when confronted with something as bizarre as dark energy. He says:

> "The real universe looks more complex than the idealized standard model that makes us infer the existence of dark energy. ... And it's presumptuous to imagine that cosmology is basically sorted out. I think we have just started."

If we extrapolate just that division of opinion between two renowned and no doubt respected academics in their field to encompass all 'divisions of scientific opinion' since the "Age of Enlightenment", what can we really claim as 'genuine' scientific breakthroughs; i.e.: That which no longer needs refinement but stands as an *absolute forever*? Not too many, one would have to say.

The April, 2009, edition of Astronomy Magazine also featured an article that offers an alternative to the Big Bang. Written by Paul J. Steinhardt: "**Why the universe had no beginning**" proposes the idea that instead of *one* Big Bang to explain the instant of creation and thus the beginning of the universe as we *believe* we know it, the *current* Big Bang was a *single* event in an *infinite* cycle. On page 33, he writes:

> "The cyclic model and the Big Bang model produce ... different pictures of the past history and future evolution of the universe. In the Big Bang view, the Big Bang marks the beginning of time, so the universe is only 13.7 billion years old. A period of inflation after the Big Bang sets the large-scale structure of the universe. Theorists introduced dark energy to explain the universe's current accelerating expansion, but otherwise it serves no needed role. Once introduced, however, dark energy dominates the future of the universe.
>
> In contrast, the cyclic view says 13.7 billion years represents only the time since the last bang and the creation of the matter and radiation we see today. The universe has had many such cycles – perhaps infinitely many – prior to the present one. And the true age of the universe is far more than 13.7 billion years.
> The theory has no need for inflation because large-scale structure derives from events that lead up to each bang. ..."

Do *we* accept the notion that the material universe into which a myriad telescopes peer is *infinite*, thus in existence *forever*? No, we do not, for the **Material World** is a "**Work of Creation**", in the same way that the **Higher Realms** are. The primary purpose of including the 'infinite cycle' theory here is to *strongly* illustrate – again *very* necessarily – the problematic nature of a scientific Discipline [cosmology] that does not possess a **'foundation of sure knowledge'** from which to conduct further research to gain greater insights. Such a 'scientific quandary' surely has resonance in **"The Error of Scientism"** previously quoted.

If science is ever to reach the point of no longer 'undermining itself' with wrong theories, it must *first recognise* what is *inherently wrong* with 'scientism' – which very few scientists seemingly want to acknowledge as problematic at all. From *that correct recognition*, then engage with and *embrace* the key 'human-entity pointers' that would facilitate the necessary transition from the present flawed state of uncertainty to one that *begins to open up* to knowledge that *mightily transcends* current astronomical *theories*.

To that end, they are:

- Our inherent free-will attribute.

- Which part of we, the human entity, possesses that free-will aspect.

- Where that free-will attribute comes from.

- How we acquired it.

- These points consequently *then lead* to the *knowledge* of the *true nature* of we, the human beings in Creation; and thus to:

- The *knowledge* of our *true* home.

In concert with those pointers, we restate perhaps the three primary "Crucial Imperatives" relevant to them. Taken from Chapter 1 [**The Crucial Imperatives**] of the Parent Work, they are:

Crucial Imperative No 2:

That we, the human beings of planet earth, are not solely a physical entity, but also <u>necessarily</u> possess a *non-material* <u>inner animating core</u>: *For the physical cannot – and therefore* **does not** *– <u>animate</u> the physical!*

Crucial Imperative No 3:

That being more than just a physical body means we naturally and *inherently* possess a <u>*separable*</u> <u>*entity*</u> **within** the material form. And that <u>*that*</u> is the *actual* life-force, the *animating* core, that is *actually* <u>*each individual*</u>!

Crucial Imperative No 5:

That because the physical Universe is a **material** expanse, it is therefore **not** without end. **It is finite!**

In the final analysis, does it really matter what astronomy postulates as the true picture of the cosmos? Our purpose on earth was always to recognise and strive to understand the **Universal Laws of Creation — Creation-Law —** by which we are enjoined to live. And to *thereby* recognise our **actual Origins** – that place in Creation from whence we came, our **true home** – and to where we are *meant* to return. So the incredible amount of time and energy expended on trying to understand the physical universe/s, never mind the huge cost of it, may be ultimately wasted – **if** those particular scientists concentrate *solely* on just the single, material aspect.

Such efforts are nonetheless invariably lauded as possessing the potential to lead to some kind of *ultimate* knowledge, which is most unfortunate. For under current educational parameters – at least in the Western world anyway – generations of students, obviously numbering in the hundreds of millions, simply follow academic lines of thought which are *ultimately detrimental* to the *more necessary* 'deeper-seeing' paradigm.

That is not to say we should not study the stars and the universe, for such investigation really is the preserve of science. The truly intuitive scientist within that particular Discipline will understand, however, that the gift of intelligence, which most scientists obviously possess, is exactly that; a gift. It is one, therefore, which should be employed for the purpose of studying and explaining the *connections between* the *transitory* material and *The Eternal* non-material; thus to *reveal and explain* the most powerful and profound *correlation* between that which is here in our *material* home-world, and that which exists *far above* the physical universes – **"the many mansions"**. As it stands today, however, science often appears to edify itself.

Science, and therefore scientists – out of themselves and ***through their work*** – should edify **The Creator** and **His Work of The Creations**. The very nature of scientific endeavour and discovery should bring about this recognition naturally in any case, particularly where the study of the cosmos is concerned. For:

> "When we consider thy heavens, the work of thy fingers,
> The moon and the stars, which thou hast ordained,
> What is man, that thou art mindful of him?"

> (The Gospel of the Essenes.
> E. B. Szekely, p 175)

Such clear wisdom from the ancient world has finally led us to the 21st century world of the Hubble Space Telescope, powerful radio telescopes and super-computers which help us better understand the make-up and behaviour of the *stars* within the 'incomprehensible' universe *we belong to*. Such wonderful aids, however, are yet still insufficient to provide final and definitive answers as to what the material universe *really is*.

The series of questions we asked regarding the impossible-to-understand numbers of the Big-Bang 'inflation model', which mainstream cosmological science evidently claims is mathematically valid, should offer you, the reader, powerful food for thought for "digesting" the next and rather mind-blowing, *key segment*, of this Booklet.

Does anyone *really* understand such 'space' distances? Notwithstanding the fact that present-day computer power can number-crunch very accurately, it all *seems* rather meaningless when many zeros are slotted behind a given digit, and perhaps to the power of another number for good measure. Nonetheless, such determined efforts to 'fix' the size of the universe has vital relevance to our understanding of the "7 Churches" question. A few quotes from various people and publications over past decades reveals the struggle to even try to begin to genuinely understand this thing known by millions of Star Trek fans, unfortunately wrongly, however, as: **Space: The Final Frontier!** [Exclamation marks!!!]

> "My suspicion is that the universe is not only queerer than we suppose, but queerer than we *can* suppose."
>
> (J.B.S. Haldane.)

> "But what came before the big bang, and how will it all end? Billions of years hence, will gravity overcome the expansion and pull matter back into its primordial state – in a big crunch? And if the universe is closed, might another big bang follow, with another expansion? Or, as many astronomers now believe, will an ever-expanding, or open, universe end in a whimper, its galaxies scattered irretrievably, their star fires spent and cold? For now, the questions are the domain of the *philosopher* as well as the astronomer."
>
> (National Geographic Star Chart, 1983.)

> "It may be that our universe is merely part of many larger universes, some in different dimensions, and that

big bangs are going on all the time all over the place. Or it may be that space and time had some *other forms altogether before* the Big Bang – forms too alien for us to imagine – and that the Big Bang represents some sort of *transition phase*, where the universe went *from a form we can't understand to one we almost can.*" ***"These are very close to religious questions."*** Dr Andrei Linde, a cosmologist at Stanford University, told The New York Times in 2001.

(All emphases mine.)

So what is the *true* nature and *extent* of the physical universes? Earth-science will surely say we may never ever know. The far greater knowledge that lies inherent in **The Spiritual Laws Of Creation** , however, absolutely states we *can* know. Moreover, it is the Spiritual Duty and Responsibility of *every* human being to *seek out* this knowledge, exactly as the great scientist Isaac Newton sought. For in that most necessary first recognition lies the far greater recognition of how far we – the human beings of planet earth in **Subsequent Creation** – *really are* from the **Very Source** of our life and *ultimate reason for being*.

0.2 The "Revelation" of 'The Plan of The World'

As stated at the beginning of the text: The key focus of this Booklet is derived primarily from the revealing knowledge of **The Book of Revelation**, for *it reveals* the wider *material environment* that the physical universes *actually represent*. Within that vast, mind-numbing "amphitheatre", therefore, the human, *physical* entity finds *its relative, contextual, place*.

As also previously stated, the vastness of the material universes is essential for the human entity to know so *we* may *recognise* that: *as incomprehensibly vast as what we will **now** reveal, the unfathomable scope and scale of it all* *is* nonetheless ***just our physical home for our physical form***! That 'immensity', therefore, *actually represents* and thus *is*, the *smallest* and *lowest* part of **Creation**.

So: What part of "The Revelation" describes that which the "Hubble" *will never see?*

For Bible readers generally, but more particularly perhaps for Bible scholars, Theologians and the religiously learned of the Universities and Colleges of the world, the question of the "Seven Churches" or "assemblies or communities" in "Asia" or "Asia Minor" in **The Book of Revelation** represents a curious and fascinating puzzle. The names of the seven "assemblies" – to each of which a messenger of God delivers an address – are given as follows: Ephesus, Smyrna, Pergamos, Thyatira, Sardis, Philapelphia and Laodicea. And, just as The Revelation ***seemingly*** notes, they *were* communities which *did* once exist in Asia Minor.

Ephesus: Ancient Greek city of Asia Minor, in what is now western Turkey, lying near the mouth of the river Ku-cuk Menderes. It was the site of the great Temple of Diana, one of the seven wonders of the world. It was destroyed by the Goths in AD 262.

Smyrna: Now called **Izmir**. City and port in western Turkey. At the head of the Gulf of Izmir, on the Aegean Sea, it is the commercial centre of the Levant.

(Great Illustrated Dictionary,
Reader's Digest; both.)

Pergamos: Now called **Bergama** in western Turkey. Once noted for its fabulous carpets; they were the most highly valued and probably woven with gold and silver thread. Nothing survives of these rich textiles because they were all burned long ago to extract the metal.

Thyatira: Now called **Akhisar**, a town in western Turkey, "... in a fertile plain on the great Zab River (the ancient Lycus). The ancient town, originally called Pelopia, was probably founded by the Lydians. It was made a Macedonian colony about 290 BC and renamed **Thyatira**. It became part of the kingdom of Pergamum in 190 BC and was an important station on the ancient Roman road from Pergamum (Bergama) to Laodicea (near Denizli). Its early Christian church appears as *one of the seven churches in the Revelation to John"*.

(Brittanica CD '97 both.
Italics mine.)

Sardis: Capital city of ancient Lydia, now a small village in western Turkey. When Lydia was absorbed into the Persian Empire following the defeat of Croseus (c. 550 BC), Sardis remained the provincial capital of Asia Minor. It later became an early centre of Christianity – *one of the seven churches of Asia (Minor)*. Extensive excavations of the site have yielded the earliest known coins, dating from c. 700 BC.

Philadelphia: Now called **Alasehir**, also a town in western Turkey.

Laodicea: Name given to several cities built in Asia and Asia Minor by the Greek Seleucid Dynasty in the third century BC The chief one, Laodicea ad Lycum, near present day Denizli in western Turkey, was a prosperous market town on the Roman trading route from the Orient and an early centre of Christianity.

(*Great Illustrated Dictionary*.   Italics mine.)

Already now we have a strong *academic* view about what the "7 Churches" *might* mean. Significantly, they were in a geographically very small area of present-day Turkey. In fact, the triangulated area encompassing the locations of those ancient "7 Churches", including the nearby island of Patmos and the area of sea between it and the mainland, amounts to something like 25,000 sq.km. The total area of modern

Turkey is around 780,580 sq.km. The representative area of the '7 Churches triangle' is therefore only about 1/30th of that.

So if we *carefully* read the introduction *prior* to the "messages" being given to the seven 'assemblies' in Asia by the individual "messengers" of God, then read the *actual* messages, and then finally the *explanation* about **The One** who has authorised the messages; a major problem immediately emerges *if* we hold to such a *small area of one small country on earth*. The island of Patmos, where it is *believed* John the Disciple 'received' The Revelation, is close by. So how do, or how might, the actual messages fit with such a scenario?

Each of the seven 'communities' is addressed by its particular messenger (a Guardian Angel) who calls the inhabitants to task for various transgressions against The Laws of Creation – The Laws of God – and warns the various 'assemblies' what will happen if they do not change their ways. Let us take just one message to one of the communities, and carefully note the **key** to understanding the *meaning* of that and the other messages.

0.3   The Vision in Patmos

> I, John,...was in the island known as Patmos. I became inspired on the Lord's day; and I heard a loud voice behind me resembling a trumpet blast saying: "What you see write in a book, and dispatch to the seven assemblies – to **Ephesus**, and to **Smyrna**" ... etc.. I accordingly turned to see the voice which spoke to me. And having turned, I observed **seven golden lampstands**; and **in the centre of the lampstands**, one like to the Son of Man... and **holding in His right hand seven stars**; and a sharp double-edged sword drawn from its sheath... "Write therefore what you have seen, what is, and what will come after these. The mystery of the **seven stars** which you saw upon my right hand, and the **seven golden lampstands**,

the seven stars are <u>messengers</u> of seven assem-
blies; and the seven lampstands <u>are</u> the seven
assemblies."

(Revelation 1:9-19, Fenton.
Emphases mine)

We have strongly emphasised the italicised parts of John's vision because they hold the key to understanding this *apparent* mystery, **for it was reserved for this point in time in humanity's journey for it to be fully understood.** Verse 3 [quoted under the Title] offers a further connection:

> "Happy are the reader and hearers of this prophecy who observe its records; for the time is at hand."

The next key part is the text of the address to the first assembly: **Ephesus**.

0.3.1 To the Assembly in Ephesus

> To the messenger to the assembly in Ephesus write: "Thus says the Controller of the seven stars by His right hand; who walks in **the centre** of the **seven golden lampstands**; I know your position, your industry, and your patience; and that you cannot endure those who are wicked; that you have put to the test those who have called themselves Apostles, and are not, and have found them false; and you have had patience and have suffered because of My Name, and have not failed. I have, however, a charge against you - that you have *forsaken your first love*! Remember, therefore, from where you have *fallen*, and repent, and practise your former works; failing which, and unless you alter your mind, **I will come and remove your <u>lampstand</u> from its place.**"[7]

Revelation 2:1-6, Fenton.
Emphases mine.

---

[7]The italicised sentences in the whole discourse are vital for the reader to remember.

"From whence thou art fallen" refers to the [human] over-cultivated intellect, which has pushed aside the spirit and caused it to fall, so that it can no longer, as *before* the Fall of Man, do "the first works", namely, keep awake the spiritual intuitive perception, and thus maintain the connection with God.

(*A Gate Opens*, Herbert Vollmann.)

The address to Ephesus continues on for a few more sentences, then there are the further, basically similar, addresses to the *other 'assemblies' or churches.*

Now, if we accept such a scenario with regard to the "7 Churches" in what is now western Turkey in the absolute literal sense, then we might appear to have three possibilities. That in a past event [because the particular cities to which the names once belonged have either disappeared or are no longer known by their ancient names] the messengers *did* appear to the people, and *did* deliver the appropriate messages. Subsequently the cities or communities *did* disappear so one *could* believe that they did *not* "change their ways" and thus suffered destruction. However, there is no known record of such visitations or warnings to those "churches" having taken place.

We note that the name Laodicea, alone, was given to *several* cities. That is surely problematic for a "messenger" of God who is required to deliver a message of obviously great import to that particular 'community'. For messages at the behest of **The Almighty** cannot be superficial or insignificant.

That being the case, the question of *which* John received The Revelation and from **Whom** is clearly important. There is a school of thought that accepts three different individuals: John the Baptist, John the Evangelist – the beloved Disciple of Jesus associated with the Fourth Gospel – and a John the Divine; believed to be the author of The Revelation. We, however, will embrace only two.

59

If we accept the view that **"The Revelation"** was received by John the Disciple on the island of Patmos in the Aegean Sea, we must logically accept that he was sufficiently 'well-connected' to so receive such a powerful 'unearthly vision'. For such a **'Revelation'** could only come from **The One** *Enthroned* at the Height of **Creation**; thus **HE** with the greatest knowledge of **It**.

Notwithstanding the fact that 'John the Disciple' was the "beloved of Jesus", such tidings of vital import must still presuppose that only a very special individual – a greater one prepared over a long period of time perhaps – would be suitable for such a high task. Or was there a *second* John on *a Patmos*?

From Matthew 11:11, "A Gate Opens", by Herbert Vollmann; we read:

> Only one was found worthy to receive the great Revelation of past and future happenings: John the Baptist, of whom Jesus said, "Verily I say unto you, Among them that are born of women there hath not risen a greater than John the Baptist."

And from The Revelation of John, p.161, of the same Work:

> "Moreover, John received The Revelation not on the island of Patmos in the Aegean Sea during his earth-life, but *after* his earthly death – on the Isle of Patmos that lies in The Spiritual Realm, even above the Paradise of the human spirits. He passed it on to a human being on earth *who was spiritually open for it,* and who translated it into earthly words. Thus the Book with seven seals, The Revelation of John, *was handed down to us.*"

(Emphases mine.)

Here we thus note that the "mansion" of John the Baptist [the "none greater"] lies far above human origins. Hence the reason, also, why **only he**, and not an ordinary priest at the time of Jesus, could baptise **The Son of God**.

Now, if it *is* believed that *John the Disciple* received 'The Revelation' on the island of Patmos in the 'Aegean', why did he not simply travel to each of the communities to deliver the messages? He was certainly close enough, would have been well respected in the communities, and therefore probably believed. Notwithstanding historical notations in some Bibles that John was *exiled* to Patmos because of his faith and in that case could not travel *from* there, his *followers* certainly could have. For we know that the four Evangelists had many. We note that the Apostle Paul journeyed to all those "churches" to preach, so they were certainly well-established.

Given that John was an especial and faithful 'servant' who would have regarded The Revelation as a singularly-important Message from God, he would surely have found a way to deliver the 'messages' to the named communities in the region. That is, of course, if The Revelation *really was about the "7 Churches" in existence there at that time.* There would not then be the need for messengers to descend into the physical part of Creation to utter proclamations.

So question remains: Did those small cities warrant such an exalted visit, and/or is it still yet to happen? But where are the names now? As we have noted, the cities, for the most part, no longer exist in their original form. Notwithstanding the fact that it was part of the region where great religions were born and from where much recorded proclamation and prophecy is derived, such a notion nonetheless still focuses on a very small geographical area. And, moreover, on a part of the planet containing just a very small number of people.

The next question we might ask if we follow this improbable thread through is: What does it mean if it *was* fulfilled in some way, even though no record exists of powerful messengers from on high visiting and proclaiming to whole communities? Surely the word would have gotten out and been recorded by someone, even if not by those in the actual communities who, however, would have experienced what would surely have been a stupendous event. After all, only a relatively few people saw the Star of Bethlehem and the "miracles" of Jesus. Fewer still saw Him in His *other-world body* prior to ascending to **'The Father'** to become **One** again with **Him**! Yet all of that is accepted by many hundreds of millions today.

Can we not also accept that *if* such a major spiritual event *had* taken place *"seven times"* in such a small area, would not the whole of the known world have been "buzzing"? But — silence![8]

On this particular issue, then, uncertainty in 'theological treatises' and the like on the true meaning of the "Seven Churches in Asia-Minor" persists among the 'University-learned' and layman to this day. Again we ask the key question: *If* we accept the popular Christian view that John the Disciple received The Revelation on the island of Patmos but could not travel to the named "communities" by virtue of his exile there, why did not *others* close to him *deliver* the messages to the "7 Churches" after John *received* them?

We know that the Disciples were instructed by Jesus to a far greater level of knowledge than most human beings, even to *this* point in time. So having been instructed in the far greater expanse of Creation itself – *"In My Father's House there are many Mansions"* – the Disciples were well aware of impending future events because they had all been told in no uncertain terms **by The Son of God Him-**

---

[8]Historically we know that the revolt of the Jews in 66 AD and the subsequent destruction of Jerusalem by Titus and his legions in 70 AD may have been initiated in part by the belief at the time that Jesus would return *shortly after His crucifixion* as the King/Messiah to inaugurate the Millennium of Peace and defeat the Roman oppressor. That this did not happen clearly shows that *that* interpretation by the religious leaders of the day was completely wrong. Indeed, they paid very dearly for that presumption with the wholesale slaughter of almost the entire population of Jerusalem. The legacy of that kind of great error continues on in theological circles today in the ongoing uncertainty about what The Bible actually states and means about Jesus ostensibly returning and bringing with Him an Apocalyptic Judgement and great destruction around this present time.

self what would come upon humanity at the "completion of the times" – *our* **times** – and why! We can also know this if we wish to, simply because it was recorded by them, and we can read that forewarning discourse today in virtually any Bible.

However, even though the messages in 'The Revelation' were recorded "for all time", it was reserved *for a later time*, a future time – *our present time* – for its sure clarification and wider dissemination, when the true knowledge of Creation would be given to humanity as promised. Once recognised, that new knowledge could be more readily understood by present-day man through the twin Disciplines of **science and theology**. Thus, via the *relative* human dimensions of man in his *duality*, i.e., **intellect and spirit**.

Note:

[Under the perfect outworking of The Laws of Creation, a point in time is set where all cycles previously unresolved between individuals and even whole peoples would need to be 'closed off'. In order for such 'closures' to be *fully understood* by all those "peoples" at that ordained time, however, the evolutionary path of humankind *should* have reached its zenith for all by then. That *"then"*, that zenith, is this present time. Therefore, the paths of religion, philosophy, science, and even the social order of all Nations and peoples, *should* have travelled a road of unfolding enlightenment founded on the great spiritual truths – given in a carefully-guided, step-by-step process through the line of Prophets and Truth-Bringers Called from Above – to culminate in this present time where *all* was ordained to be revealed.]

Hence the need for empirical scientists and 'religious theologians', particularly, by this time to have subjugated their singularly-egoistic positions *ostensibly* providing the *primary* truth for all of humankind, and recognise that both Disciplines had an equal part to play in leading global humanity to complete and final knowledge of its true origin and purpose. Unfortunately, however, the window of opportunity for any kind of real accommodation in an harmonious and fully-knowledgeable working together to enlighten humanity has probably now closed.

And because it is primarily **untruth** that 'educating-academia' have fed to the masses; at this decisive point in humankind's journey, that task of **enlightenment** has now fallen to a **few radicals**, a few **"voices in the wilderness"**, to reveal the **errors** and thereby **proclaim for**, and thus **lead to**: **The All-Truth!**

So, if the messages were not actually for those ancient communities, what might it all mean? In order to understand this great question, a related one stands out as also being particularly important. It is the necessary understanding of where we, as human beings on planet earth in the World of Matter, *actually stand* in relation to: **The Creator!**

> For **He** surely cannot stand in the same part of **His Creation** as we do, i.e. as a Presence or Power **actually residing <u>in</u> the physical universes**. **And neither does He!**

Yet there should be an inherent wish on our part *to want to know* where we do *actually* stand! Do we, upon earthly death – as so many *apparently* believe – simply transit from earth and be immediately in 'paradise' and thus in the presence of God? That seems to be the general, broad belief of at least the three major monotheistic religions. We, however, from our broader analyses in the Parent Work, already know the answer to that question. For if it *was* possible to be in **The Almighty's** immediate presence; if it *could* be so, then that would very *illogically* place **Him** quite close to we human beings of earth. Yet our critical examination of The Book of Genesis and the *two Creations* in the Parent Work clearly

64

negates such a view of a close and convenient, personalised God. ***Simple logic should tell us that anyway.***

A connection between the '7 assemblies' and where we stand in relation to The Almighty might not be immediately apparent, or even seem logical, but the ***true understanding*** of what the "Seven Churches" ***actually means*** offers *precisely* that *correct knowledge and relationship*. A *correct interpretation* tells us what ***they*** (the "Seven Churches" or assemblies) actually are as a ***complete***, ***overall entity***, and what ***it*** (each assembly) is in its ***singular***, ***individual*** form.

We have determined that this "mystery" is not solely a religious question only, and that *certain* Disciplines within the 'scientific communities' as well need to think long and hard about what it all might actually mean. Since we contend that purely intellectual seeking alone – from either of the Disciplines of science or theology – will not permit any unveiling of the *true* meaning of the "7 Churches in Asia", we need to therefore take the boldest possible step in order to "find". By *merging* the two Disciplines and using the clarified intellect guided by the spirit – as we have previously done with other key questions in the Parent Work – we will achieve exactly that goal.

We will, thereby, indeed discover a truly marvellous revelation; **the very revelation that Newton himself sought!** A revelation which is literally mind-blowing in its ramifications for science and for ***all*** religions, permitting the mind, intellect, soul and spirit to soar in exultant recognition.

> Therefore: *If recognised and understood correctly,* the senses will ***reel*** before the ***true meaning*** of **"The Seven Churches in Asia-Minor"**. For it ***completely 'shatters'*** all current ideas deriving from astronomy and cosmology about ***the true nature*** and ***extent*** of the 'universe/s'.

It also reveals therewith the huge importance of 'The Revelation' itself. The almost incomprehensible import of such a vastly overpowering, yet *spiritually-empowering,* extra-world view of where we human beings **actually stand** in Creation is virtually life-changing in that singular moment of *genuine* **recognition.**

To that end, the intuitive insights of **Pope Benedict XVI**, **Einstein** and even **Galileo** – we already have **Newton** with us – are exactly appropriate for this Booklet revealing 'Isaac Newton's' "Plan of The World". More especially for singularly-focussed and perhaps hard-nosed cosmologists who might cross the path of this Essay, the following *crucial insights* of one noted *theologian* and three of the **greatest foundational-scientists ever** offer an especial help for *present-day* cosmologists on **how** to understand **"The Revelation of the Cosmos"** — herein!

So, between the three great *foundational space scientists* – never mind their many *other* talents – we firstly have **Galileo** re-anchoring and thus *rescuing* the truth of the **Planetary System** from an egotistical and tyrannical 'religion' masquerading as the sole voice and representative of the profound and sublime Truth brought *down* to the earth by **The Son of God**. Yet a Church which nonetheless cynically distorted **His Truth** for earthly ends.

Newton's seminal discoveries emerge as *especially crucial* to the furtherance of 'space knowledge'. His **"Law of Gravitation"** allowed for further discoveries of great import. And, of course, **Einstein**; believed by some to be the greatest mind of all. His **"Theory of Relativity"** gave space-science especial keys to understanding much more about the cosmos than was previously possible.

However, whilst these *undeniably great scientists* ostensibly worked *solely* under the ethos of scientific-empiricism for their time, *present-day* scientists really need to recognise the clear fact that those three *especial men* obviously stood *way outside* the 'square box' of their contemporaries. For by virtue of the fact that they *were* so far ahead of their 'peers' – apart from the necessary empirical aspect of mathematics to prove their 'discoveries' – intuitional insight must have played a very crucial part in allowing them *in the first place* to *see* or *perceive* what others could not. It could be said, therefore, that their seminal discoveries and associated work *almost* placed them in the "lone voices in the wilderness" category.

From that necessary foundation and the amazing discoveries since; should we now say?:

Space! The <u>Final</u> Frontier!!! — [Exclamation mark/s!]

Or; as Newton has insightfully intimated *for* present-day cosmologists and 'space-science' and with which we unequivocally concur; should it rather be?:

Space? The Final <u>Frontier</u>??? — [Question mark/s?]

Specifically in concert with the great intuitive perception of Newton, and perhaps even *high spiritual guidance* in *his* case, but also with Pope Benedict's spiritual insight; the final and greatest *space discovery* of all — **Sir Isaac Newton's "Plan of The World"** — we now reveal!

Whether believed or not, space scientists should at least be *intrigued enough* to *wonder* at what Newton was *really* searching for. Given the very radical nature of what will be unveiled, it is important to state the nonetheless accepted fact that Newton **was not a deluded fool**. And no scientist today would dare label him deluded, at least not publicly.

So: Of the *two* possibilities about 'space' – and only *one* can be right – the *correct* position and thus the *greatest knowledge of all* about the **"cosmos"** equates to:

Space? The Final <u>Frontier</u>??? — [Question mark/s?]

> "Modern scientific reason quite simply has to accept the rational structure of matter and the correspondence between *our spirit and the prevailing rational structures of nature as a given*, on which its methodology has to be based. Yet the question *why* this has to be so, is a *real* question, and one which has to be remanded *by the natural sciences* to

67

other modes and planes of thought – to *philosophy*
and *theology*."

(Pope Benedict XVI. Regensburg, 2004)

"Though religion may be that which determines the
goal, it has, nevertheless learned from science, in the
broadest sense, what means will contribute to the at-
tainment of the goals it has set up. *But science can
only be created by those who are* thoroughly imbued
with the aspiration toward **truth** *and* **understanding.**"

*"Science without religion is lame, religion without sci-
ence is blind."*

(Einstein. Ideas and Opinions, p.42-3)

On the relationship between science and religion, Einstein
notes:

"Intelligence makes clear to us the interrelation of means
and ends. But mere thinking *cannot give us* a sense of
the ultimate and fundamental ends. To make clear
these fundamental ends and valuations, and to set
them fast in the emotional life of the individual, seems
to me precisely the most important function which religion has to perform in the social life of man. And if
one asks whence derives the *authority* of such fundamental ends, since they cannot be stated and justified
merely by reason, one can only answer: they come into
being *not through demonstration* **but through revelation**, *through the* medium *of powerful personalities.*
One must not attempt to justify them, but rather **to
sense their nature simply and clearly.**"

(All emphases mine.)

Though a very great mind in both the intuitive and intellectual sense, Einstein's sometimes almost 'tongue-in-cheek'
quips nonetheless hold profound truths for a *perceptive* reader.
For example:

"I never came upon any of my discoveries through the process of *rational thinking*."

Galileo: Regarded by some as the first true scientist. Upon his publicly stated recognition that the Planetary Model Copernicus proposed was correct, the Inquisition forced him to recant. He is said to have muttered under his breath, however: "But it [the Earth] does move."

Noting the Church's obsession to have Bible Scripture support Ptolemy's erroneous view of the Planetary System; anecdotally, Galileo evidently thought that whilst The Bible *was* 'the Word of God':

"...it was not a good astronomy book".

The huge strides made in astronomy since his death in 1642 would surely have astounded him – as it does modern man. However, notwithstanding all the marvellous discoveries since, Galileo was not correct in believing that astronomy was not in The Bible. As the overall thrust of the Parent Work unequivocally states:

The Bible is a Primary Work of Foundational-science!

It was thus left to **Newton** to *seed in the minds of **modern** astronomers* the **golden key** to discovering the *final revelation* of the **true nature** of the cosmos. However, as the great man Einstein intimates, rational thinking will not do the job. Revelation and *spiritual intuition*, however, in concert with a *clarified intellect*, most certainly will.

As previously stated, since we have addressed the other key questions from a fundamentally different and far-reaching paradigm, this particular one, too, can only be logically understood by using the same method. However, because The Book of Revelation primarily reveals insights into vast Spiritual vistas and events that must necessarily *transcend* intellectual interpretations; so, similarly – and by a considerable margin – must we also *vastly expand our human frames of reference* to answer and understand the *true meaning* of the "7 Churches in Asia Minor".

We contend, absolutely, that the premise we postulate is correct; that the question under scrutiny here is **not at all about earthly communities or churches**. The fact that those churches did once exist has clearly confused the issue.

Where, then, might these places be? No satisfactory answer has publicly emerged from the Jewish faith, the Christian religions or the institutions of Theology. In any case it is not about guesswork or theories, but about the knowledge of the outworking of **The Laws of Creation** and the concomitant knowledge of the **structure** of **The Creation** that has issued from the very outworking of those immutable Laws. Therefore, there *really is* a correct answer; there is *always* a correct answer! So what does that imply?

In summary, since we have concluded that the answer is **not connected** with the names of the "Seven Churches" in the stated general location of Asia-Minor on *earth* in ancient times, the so-named 'assemblies' must therefore represent something **entirely different**. That logically places *both* the communities *and* the delivery of each specific *address* by a "messenger of The Lord" [the Guardian Angel of each assembly] **somewhere other than** the earth, in a time **different to the times** when those *named* communities once existed.

Thus far we have sparred with possibilities and ideas about the meaning of the "7 Churches or assemblies in Asia-Minor". If, as we unequivocally state, the 'assemblies' are *not* on earth – but nonetheless still exist – they would therefore **have to be**, quite logically, in **The World of Matter**. **For this is where "The Revelation" came *down* to; where we *live* in our obvious *physicality*.**

However, a seamless transition from current uncertain views to sure certainty in an instantly-recognisable answer is not 'a simple fix' in this case, for it requires a huge and fundamental leap of truly gigantic proportions *into a completely new paradigm.*

Moreover, it is not a paradigm that is simply and solely religious in nature and import. In reinforcement for this particular "revelation", *it also encompasses astronomy, astrophysics and cosmology on the most stupendous scale.*

Most unfortunately, however, a major problem preventing a wider outlook in the present lies with the so-called "academic elite" from the world of astronomy/cosmology. For *they* are the "experts" in their field, and only *they* are supposed to know. Notwithstanding how incomprehensibly vast the 'known universe' is, what stifles any wider vista is the general *inability* of cosmologists to recognise that there is *much, much more to Creation than the seemingly-endless vault of our star-studded night sky* – itself a tiny part of just our home galaxy, never mind the many more billions we are now aware of.

Even with these expansive discoveries, however; because 'space-science' teaches a *smaller* Material World than is *actually the case*, by educational default *the greater mass of humanity must also accept such a view.*

Today, with radio telescopes and the Hubble space telescope, we possess the ability to peer far further into the great expanse we euphemistically label 'space' than was ever thought possible. Yet even with those marvels of engineering and inventiveness gifting us truly marvellous and stunning images of ever new discoveries of the cosmos, *all of that does not even **begin** to get **close** to the **actual** nature and **size** of just the Material Part of Creation.*

*Astronomers and cosmologists, **as a global group**, do not yet **know** this to be the case.*

In an early Chapter of the Parent Work, we unequivocally stated that Creation and Evolution are necessarily one and the same. In reality, there is no separation. The old rivalries between science and religion are just that – rivalries.

Rivalries based on the narrow parameters that each side promotes for its own edification. With regard to the question this Booklet examines, there is no actual separation there, either. Spiritual truth and earthly science were ordained to be *mutually inclusive under the parameters of **The Laws of Creation.***

With the huge strides made in the 'science' of astronomy and cosmology wrought by the building and use of more and more powerful instruments to observe the great expanse *"out there"*, a fascinating paradox has seemingly emerged. It would appear that because the expanse of the, now, known universe is simply too incomprehensible to grasp, a point or limit in our brain-capacity to truly understand would inevitably be reached. So that any greater understanding would perhaps derive initially from the conjoined and expanding paths of philosophy and religion, but only reaching a complete and final picture with, and from, the knowledge of the actual "**Structure of Creation**" of which the *material* universes are, to emphasise once again, *the lowest and smallest part*.

0.4   The 'Meaning' of the "7 Churches"

If we now move a little closer to understanding the meaning of the "7 Churches" question proper, we should revisit the notion of what might happen if we *could* travel to the edge of the universe and look beyond it. To what, we may wonder?

> According to Einstein's Theory of Relativity, man could never reach that point because "...the universe bends in a way that can't adequately be imagined". Therefore we would, even after travelling in a straight line to the edge, eventually arrive back at our starting point because space curves in a way that allows it to be boundless – *but finite!*[9]

---

[9]Here is the key understanding to the answer of the "7 Churches" question. The physical universe we see and believe we know is *finite!*(Crucial Imperative No. 5 in Parent Work.)

"In short, there's more space than you can imagine already without going to the trouble of trying to envision some additional beyond."

(*How to Make a Universe.*
Emphasis mine.)

That is *exactly* the purpose of this Booklet, however; to *explain* that *'additional beyond'*. Therefore, since the science of astronomy/cosmology has seemingly not concerned itself with the great and key truths contained in certain religious writings, let us use the Spiritual/intuitive faculty available to the human entity to now finally explain and clarify – *for science and theology* – the true meaning of "**The Seven Churches in Asia-Minor**".

To do that we need to once more consult, absolutely fittingly in this case, **The Book of <u>Revelation</u>**, 1:9-19. In particular "The Vision in Patmos". (Emphases mine.)

The Vision in 'Patmos'

I, John, ...was in the island known as Patmos. I became inspired on the Lord's day; and I heard a loud voice behind me resembling a trumpet blast saying: "What you see write in a book, and dispatch to the seven assemblies – to **Ephesus**, and to **Smyrna**"... etc.. I accordingly turned to see the voice which spoke to me. And having turned, I observed **seven golden lampstands**; and **in the centre of the lampstands**, one like to the Son of Man... and **holding in His right hand seven stars**; and a sharp double-edged sword drawn from its sheath... "Write therefore what you have seen, what is, and what will come after these. The mystery of the **seven stars** which you saw upon my right hand, and the **seven golden lampstands**, *the seven stars are <u>messengers</u> of seven assemblies; and the seven lampstands <u>are</u> the seven assemblies!*"

The key to understanding this *apparent* 'mystery' again lies mainly in the *italicised* words. The next key part is the text of the address to the first 'assembly' of **Ephesus**.

> To the messenger to the assembly in Ephesus write:
> Thus says the Controller of the seven stars by His right hand; who walks in **the centre** of the ***seven golden lampstands***; "I know your position, your industry, and your patience; and that you cannot endure those who are wicked; that you have put to the test those who have called themselves Apostles, and are not, and have found them false; and you have had patience and have suffered because of My Name, and have not failed. I have, however, a charge against you – that you have forsaken your first love! Remember, therefore, from where you have fallen, and repent, and practise your former works; failing which, and unless you alter your mind, ***I will come and remove your lampstand from its place***."

> (Revelation 2:1-6, Fenton Bible.
> Emphases mine.)

Here, too, the *italicised* sentences represent a *golden key* to the full and final understanding of the meaning of the "7 Churches in Asia". And thus – in relation to the vast gulf and ***humanly-unbridgeable distance*** between we of Subsequent Creation and **"The Creator"** – where we *actually stand* during the *earthly phases* of our complete existence.

We note that the address to Ephesus continues on for a few more sentences, then basically similar addresses are given to the other 'assemblies' or 'churches'. Notwithstanding the unequivocal stance we are postulating here, we nonetheless clearly recognise that the Messages *as they stand* certainly give the *impression* that they were/are addressed to the seven ancient cities of Ephesus, Smyrna, Pergamos, Thyatira, Sardis, Philadelphia and Laodicea.

For words like Jews, synagogue and Satan are used, along with the phrase, 'sons of Israel'. Moreover, we have mention

74

of Balaam and Balak, names associated with ancient Asia Minor. Luther also cites the Nicolaitans, who are supposedly corrupting two of the "7 Churches". It appears, however, that no one has really confirmed, historically, who they were.

Now, since we are revealing the *actual reality* of that which Isaac Newton assiduously searched for – which must necessarily and inherently encompass a far greater expanse than the present, nonetheless mind-boggling, 'vista' currently accepted by science and Christendom – we therefore need to similarly understand that those particular and well-known terms used in the addresses must also relate to that *comprehensively* wider view.

And because The Book of Revelation possesses a powerful *spiritual component* that reaches *way beyond* **The World of Matter** – the size and extent of which we 'earth-humans' will never ever be able to truly grasp – **Isaac Newton's Plan of The World**; the words used in the seven addresses thus *similarly possess* a far deeper *spiritual* meaning.

Thus, a 'Jew' here simply means 'a spiritual person', synagogue akin to a 'place/assembly of worship/ers'; and 'sons of Israel', a 'righteous group'. Satan – or the 'Being' Lucifer – affects the *whole* of The World of Matter, not just we on earth.

By way of example, if we look at a small segment of the addresses to Smyrna, Pergamos and Philadelphia – respectively;

> "...and the insolence of those who assert themselves to be *Jews* and are not, but are, on the contrary, a *synagogue* of *Satan*."

> "...because you have there some who are holders of the teachings of Balaam, who instructed Balak to place a stumbling-block before the *sons of Israel*,..."

> "Therefore, I will give those of the *synagogue* of *Satan*, who assert themselves to be *Jews*, and are not, but lie;..."

> (Revelation, Fenton Bible.
> Emphases mine.)

So for humankind on Planet Earth in the World Community of Ephesus; our *ostensible cosmic reality* is that we live in a physical environment on a small and beautiful planet in an expanse so vast as to be totally incomprehensible. Because it *is* so vast, the obvious *assumption* that *could* be drawn is that the *visible* universe really is the *be-all* and *end-all* to our total existence.

The Material World, however, is purely for the purpose of developing to personal self-consciousness and maturation, both intellectually and spiritually. Whilst a good balance between those two aspects is essential for an harmonious working toward achieving necessary earthly goals, the *greater* responsibility of man was always to develop the Spiritual part within to the highest possible level attainable. And thereby *lead* the intellectual part.

A larger emphasis placed on developing the Spiritual part would have thus ensured that the connection to, and recognition of, our actual Spiritual Origins would *not* have been lost, and we would not need to struggle to understand the key questions to life such as the one we address here. Therefore, the Material Worlds designated for the creature, **man**, to mature in should not be assumed to be just that which we *believe we know* and perhaps even *roughly* understand. For that would, indeed, be arrogance. From an extremely narrow, solely earth-oriented, viewpoint, earth-science prides itself in believing that it, alone, possesses the ability to solve the ultimate questions to 'life'.

Such a stance, however, represents a "delusion of denial", where the very sciences themselves become a kind of all-knowing god for their adherents. Such a belief logically states that there cannot be any other avenue open to mankind for those very answers which science believes *only it* can answer. To that end, let us *once more* revisit and perhaps brutally-reinforce a key truth.

> "Scientism, the aura of authority carried by scientists, has made us believe that knowledge obtained by scien-

tists is the ultimate authority, that as we accumulate information, our capacity to understand, control and manage our surroundings will grow correspondingly. But the basic principle of scientific exploration contradicts this faith: knowledge comes from empirical observations, which are "made sense of" by hypotheses, which in turn can be experimentally tested. All information is open to being disproved. As Jonathan Marks has pointed out":

"...the vast majority of ideas that most scientists have ever had have been wrong. They have been refuted; they have been disposed of. Further, at any point in time, most ideas proposed by most scientists will ultimately be refuted and disposed of... Science, in other words, undermines scientism."

(Dr David Suzuki. *The Sacred Balance*, p.19)

If, then, we are able to accept even just the *possibility* of Realms or regions that are *non-material* – the "many mansions" – and which therefore lie *far above* the *physical* universes, then we gift to ourselves the recognitory-acceptance that the physical-world part of Creation is, very logically, **far larger** than we could **ever** have supposed. By that we do not mean just what we now know exists in *our* universe, but that a far greater reality *really does exist*. And, moreover, that the key to this stupendous knowledge has been with us **for a very long time**.

Since the 'Material World' that we know is the *physical* home of man, why should we assume that the *entity*, **man**, resides *solely* on planet Earth *in our particular universe*, and therefore nowhere else in the *complete physical expanse* we allude to? Such a view assumes that we know and understand more than The Creative Power Which permitted us conscious life. By such a belief, we actually limit ourselves to very narrow parameters of *so-called* great knowledge.

So, if those messengers of God did *not* address seven *earthly* communities of the ancient world, yet nevertheless did, or will, address *'communities or assemblies of men'* upon

77

the directive of the **"One like unto The Son of Man"**, that surely reveals the stupendous scale of the actual happening. A happening, moreover, *that could not possibly be confined to a very small part of one very small region of one very small planet.*

Accepting that *that* is the actual reality of things here, we shall now simply take the boldest step and proclaim, here and now, *that Truth* which John, the receiver of **The Revelation** from a **Higher Sphere**, understood and passed on down to humankind *primarily for this present time.*

The very truth that he received came from **The One by Whom** we came into being – **IMANUEL** – **The One Enthroned** – **The One Who Comes: The WILL Of GOD!**

He It was **Who** brought into being *the "seven homes of humanity"*: — *the "7 Churches".* It is **He**, therefore, **Who**, [figuratively], *stands in the centre of the seven golden lampstands.*

And thus we note; "...**seven golden lampstands**; and in **the centre** of the lampstands, one like unto the **Son of Man**... and holding in **His** right hand *seven stars*".

So we are not only given *indications* of what the *seven golden lampstands* actually are, but also the *key* to the *meaning* of the *seven stars.*

Thus: "The mystery of the *seven stars* which you saw *upon my right hand*, and the *seven golden lampstands*, the *seven stars* are '**messengers**' of *seven assemblies*; and the *seven lampstands* <u>are</u> the *seven assemblies.*"

(All emphases mine.)

There lies the answer.

The *seven 'assemblies'*, which *are* the *'Seven Churches' in 'Asia Minor'* – in each of which *resides a "community of men"* – *are* the *seven golden lampstands* of *The Revelation.*

They are actually *seven universes* of roughly the *same size and configuration* as the *universe* in which <u>we</u> *reside. And in each an 'assembly' of men; of humanity.* We of planet earth have our *physical home* in the *'universe'* or *'church' called* **Ephesus!**

The reader must clearly understand that these "lampstands" are not simply seven universes *within* the huge expanse of the cosmos visible to us via our radio and optical telescopes. No! Each universe or "church" is an incomprehensibly-vast rotating 'island', *each a separate entity unto itself*: Each one *similar* to our own in mind-numbing size.

The seven universes together *rotate* in a huge *wreath-like formation* at the *lowest part of Creation.* The distance *between* each universe is naturally *greater than* the *diameter* of *each individual one.* We will therefore never be able to see the other universes and, quite obviously, never be able to travel to them, at least not by physical means.[10]

The "Revelation", however, permits us to *know* about the other '6 Churches'. The designation, "Asia-Minor" – used to describe the 'Material-Creation area' of the "7 Churches" – is simply a Spiritual one, as are the names of those '7 Churches'. Thus the "mystery" of the "7 Churches" is not a mystery at all.

The Bible — A Primary Book of 'Foundational-Science' — provides the <u>cosmological</u> answer!

That is precisely why such a fundamental leap into a far greater level of knowledge was reserved for this particular time in our evolutionary development. For it is only within the last decades that we have had the ability to see our blue

---

[10]The same constraint applies to just our own galaxy, so it is not likely we will ever even journey to some of our "closest" star-system "neighbours". In any case, what would be the point of trying? Our true home is not "down here"!

planet in its wholeness from a point outside of it; a picture free of all the superstition and ignorance of the past.

As David Suzuki points out:

> "We have to recall the image of the planet from outer space; a single entity in which air, water, and continents are interconnected. That is our [physical] home."
> (Parenthetic addition mine.)

Just as that sublime vision has produced awe in many, so were we, via such technological aids as the Hubble space telescope, for example, meant to take cosmology to *new heights* in the recognition of the awesome greatness of just the vastness of the physical universe *alone*. The sheer and unfolding scale of it should thus have produced the *certain* recognition that what was being revealed, in the *physical sense*, could yet only represent the heaviest and therefore *lowest and smallest part* of Creation. [**Newton's Law of Gravitation!**] And that there was very much more beyond that 'small immensity'.[11]

That should have been a watershed recognition for humankind – especially cosmological science – recognising, *in humility*, that we are only 'developed beings', far from our Creator. Yet even though we stand a very, very long way from The Source of Life, Its ever-present Grace permits us the *possibility* to eventually *leave behind* the confines of the material parts of Creation to return to our *true home* – our point of Origin in a far Higher Realm.

Since the thrust of this Booklet *can* offer ***seriously awake people*** explanations and answers to at least one of the major questions of life, the rapidly deteriorating state of humanity today producing such a thing as an 'Apocalyptic Judgement' scenario might inevitably surface. If so, a very telling *human* paradigm/question arises here.

---

[11] Had man not fallen "spiritually", the same degree of knowledge and insight would still have been his – and very much more besides – to enhance the mechanical and optical aids we have today.

'Why is it that we of the human race — who so desperately want to be rid of the bloodshed, violence and evil that permeates global societies and cultures with increasing ferocity; who long for a world where abject poverty and human exploitation is eliminated; and for a world that will not end up the nightmare we **all know** we are rushing headlong towards — obtusely refuse to accept the *only* mechanism that *can* bring about precisely that "better world"?'   (Author.)

And the answer?

'Quite simply: The *only* way that this desirable state *can* be achieved is through the **complete destruction** of the **very many** who **are** the problem!'   (Author.)

The overall 'message' of the Parent Work clearly explains that humankind was given a long period of time in which to learn and experience all that was necessary for the recognition of Spiritual Truth – the *primary* human responsibility. Refusal to do so logically presupposes that within *immutable* **Creation-Law** parameters – which ultimately govern *all* processes, events and outcomes – there must come the inevitable day when 'the ferryman must be paid'. A period of 'cleansing' with apocalyptic accompaniment for those who refuse to accept and/or recognise such Truth, therefore, must logically reveal a cut-off point preceding any kind of catastrophic 'Judgement-phase' for the intransigent of planet earth.

Such a 'cut-off point' obviously represents the moment where there is no further time left to *redeem* or *put right* previous transgressions. That opens the way for the cleansing or "sorting-out" phase. After the "sorting-out" process comes the next phase – a time of harmonious peace no longer marred or harmed by what went before, i.e., that which *needed* to be *excised* and *removed*.

What happens, however, if the collective transgressions of a given 'assembly' of humanity in the *wider Material World* are so serious that there are no redeeming points whatever? What, then? If we look at the message given for us here in Ephesus, it is very chilling in its clear import:

> "Thus says the Controller of the seven stars by His right hand; who walks in **the centre** of the **seven golden lampstands**; I know your position, your industry, and your patience; ... I have, however, **a charge against you** – that you have *forsaken your first love*! Remember, therefore, from where you have *fallen*, and *repent*, and practise your *former works*; failing which, and unless you alter your mind, **I will come and remove your <u>lampstand</u> from its place."**

Revelation 2:1-6, Fenton.
Emphases mine.

Remember that **each lampstand** is actually **one of the 7 great universes**, in which an "assembly of men" resides.

Now, since the Material World encompassing the "7 Churches" or 'universes' is the *lowest* and thus *smallest* part of **Creation**, a removal of one **lampstand** or universe – as vast as we recognise it is – must nonetheless be understood to be a completely lawful and *simple* process for the *humanly-incomprehensible* Power which *produced them* and very, very much more besides. For the 'assembly of men' in each, however, destruction on an unimaginable and truly apocalyptic scale which, however, is not imminent for Ephesus at this particular time on that gigantic scale.

What is stated here is the stupendous, constantly-recurring, 'renewal-cycle' of birth, life and disintegration that *all* the 7 great universes or assemblies – since the moment of Creation – must now undergo in a time-frame of cosmic proportions that could not ever be even imagined. Thus *finite* universes rotating through an *infinite* cycle of birth, life and disintegration.

We who reside on the tiny speck of the earth in the great 'universe-community' of Ephesus; do we even begin to recognise from whence we have fallen? Unfortunately we do not. More unfortunately, truly Spiritual aspirations are often derided and dismissed as irrelevant and of no consequence or importance for 21st century scientific/intellectual man. They were not considered all that relevant last century either.

And what was our first love with which we no longer bother? The love for **The Creator**, as stated in **The First Commandment**, or the love for one's neighbour? Not much evidence of either in today's world. There is, however, much love and adulation for our much-vaunted, totally aspiritual intellect, precisely by which we *fail* to practise our former works – the far earlier spiritual ones. Carefully and craftily nurtured by the present intellectually-derived, university-driven 'education system', only the adulation of *human* 'works' and 'achievements' *by humans* holds sway today. Even within global religions, the 'love of man' is far greater than **Love of The Creator!**

Clearly, then, our former works – which we are directed to once more practise – were better than those we produce now, *before* our spiritual fall and decline. Failing which, and unless we change our ways, our ***lampstand,*** the universe containing our earthly home, may well be removed. In other words total disintegration of the ***whole universe of Ephesus*** at some designated future point. What is this, just religious ranting? Or is it the clearest warning yet that we could possibly receive to change our ways and begin living correctly according to **The True Laws of Life: The Spiritual Laws of Creation? — Creation-Law!**[12]

The degraded and poisoned state of planet earth today, along with the dysfunctional social order of most societies

---

[12]If science truly wishes to understand the immutable interconnectedness of **Creation-Law** in its *inviolable* reality, it must engage with and embrace the immutable knowledge of those "**Laws Of Creation**". [Chapter 3 of the Parent Work. See end of Booklet.]

globally, clearly testifies to totally wrong thinking and wrong practices by we of the 'human assembly' of Ephesus.

Since we have now reached a critical situation in just scientific terms – let alone any stronger effect from more powerful forces – it may only be a relatively brief matter of time before the 'tipping-point' that earth-science states is very close now, adds its particular measure to our cup that already "well and truly overfloweth". Are we presently rolling towards that cut-off point at a rapidly accelerating pace? Judging from world events, it would certainly seem so. Faster and faster to our self-willed destiny of 'great destruction' perhaps. And maybe survival for a few small groups and individuals scattered around the globe.

In the same way that The Bible has permitted us the knowledge of the *extent* of the physical universes in the Material World, might we also find within that Work the same kinds of clear revelations with which we can understand the actual **Why** of our poisoned and degraded earth? And thereby further reinforce our unequivocal premise that The Bible really is: **"A Primary Book of Foundational-science!"**

One especial Prophet of the Old Testament saw in visions what has already begun to happen. Isaiah, often referred to as the Great Prophet, proclaimed the kinds of events that have now not only occurred, but the outcome of which has forced earth-scientists to rewrite their 'theories' about the 'power' of earth activity. For man's evolutionary journey to *true* knowledge, Isaiah saw in his time what we today know to be 'activity' associated with 'Plate Tectonics'.
From Isaiah 24:19-20.

> 'The "Earth" is utterly broken down, the "Earth" is clean dissolved, the "Earth" is *moved exceedingly.*
>
> The "Earth" shall reel *to and fro* like a drunkard, and shall be *removed from her place.*'

How much of that chilling potential can we attribute to human decisions, societal attitudes and concomitant global activity? Isaiah certainly pulled no punches in identifying why. Bad human decisions will obviously produce bad outcomes, not only for human societies but for the planet as a whole. That is our present legacy and reality.

Decisions = consequences!

> 'The "Earth" mourns and fades away, the "World" languishes and fades away, the haughty people of the "Earth" do languish.
>
> The "Earth" also is defiled under the inhabitants thereof; because they have *transgressed* the "Laws", *changed* the decrees, *broken* the everlasting covenant.
>
> Therefore has the curse devoured the "Earth", and those that dwell therein are desolate: therefore the inhabitants of the "Earth" are burned, and few men left.'

(Isaiah 24: 4-6. All italics mine.)

What might modern-day scientists and 'empirical science' make of the above statements? Except for the relatively few who have broken out of the strait-jacket of 'empirical intellectualism', most would probably dismiss the whole idea of Bible prophecy – accurate enough to pinpoint future events of such catastrophic power sufficient to produce a "reeling earth" – as religious rubbish. Ferrar Fenton's translation of just one line of Isaiah's vision [24:20] —

"And the earth's foundations shake!"

— should strike a chord of serious warning for all scientists whose field of 'expertise' lies in anything even remotely connected with the earth.

So: What does this *really mean* for the earth-scientist? More to the point, perhaps, should scientists take note of such prophecies, or be derisively dismissive of them? As we have already stated, what Isaiah was gifted to 'see' and give warning for future humanity was the ***vastly amplified activity*** of **'Plate Tectonics'**.

When first mooted early last century, it was, of course, dismissed by the 'scientific establishment' as impossible nonsense. Now we all know **Plate Tectonics** to be "scientific fact". Surely a perfect example of **"The Error of Scientism"**. A "reeling earth" which 'shakes its very foundation'? Is this impossible nonsense too? Not now! For this we also now know to be "scientifically feasible". How?

Earth-science and global humanity have the quite recent experience of knowing that *the whole earth* can actually shake from just one, albeit very large but nevertheless, ***single***, earth movement. The devastating and catastrophic experience of the Indian Ocean tsunami, Boxing Day, 2004, has surely shaken any scientific scepticism that such a thing was impossible.

In any case, what has *already occurred* a number of times in the past with obviously catastrophic consequences for the earth could well happen again — a truly cataclysmic 'pole shift'. Such an event would *absolutely* fulfil Isaiah's vision.

A "reeling earth" with mountains and islands 'moved out of their places' would certainly result from a 'pole shift'; with, of course, unimaginable and horrendous death and destruction. A similar, though perhaps *less destructive* outcome could also occur with multiple, simultaneous earth-movements, or as a very rapid 'sundering' ripple-effect. The destruction and devastation wrought by earth movements on that scale would produce tens, if not hundreds, of millions of lives lost. So the precedent for Isaiah's "shaking earth" has now been experienced by 'science' and present-day humanity!

Should we hope that the Great Prophet's *harder vision* will not eventuate? Or has human behaviour and activity already proceeded too far past what is acceptable to The Laws of Life ..."because they [*we*] have ***transgressed*** the "Laws", ***changed*** the decrees, ***broken*** the everlasting covenant."

The Parent Work strongly states **The Bible** to be "**A Primary Book of Foundational-science**". From out of its **Book Of Revelation**, this Booklet has derived the cosmological answer to the meaning of "**The Seven Churches in Asia-Minor**". The clear visions of Isaiah depicting Plate Tectonic activity sufficiently strong to 'move the earth' gives one more exceptionally clear and *scientifically-proven example* of the 'scientific validity' of **The Bible**.

The path we have traced, explored, analysed and explained in this Booklet marries the earth-science of astronomy/cosmology to Bible Revelation and theology in a way that both transcends, illuminates and conjoins the respective views to provide the sure answer to the "**7 Churches**" question: Isaac Newton's "**Plan Of The World!**"

Therewith may we now know our 'human' place in Material Creation, and thereby *perhaps* to *begin* to perceive **The Immutable Truth** of the vast, *humanly-unbridgeable distance* between us and **The Creator**!

0.5   An Astronomy Lecturer's Recognition

In order to more fully understand what has been revealed in this Booklet, I recommend that the lay-reader, particularly, obtain a reasonably comprehensive star-chart of the universe showing its expanding nature from the Solar System to the galaxies, to groupings of galaxies and so on – to the known limits of the universe. [National Geographic have produced excellent charts over the years.] Then, using any *mainstream* Bible, apply the texts used herein to a visual examination of the star-chart and recognise the immensity of just our universe of Ephesus to begin with.

Then travel beyond that *in mind and spirit* into the area of Material Creation occupied by the other "6 Churches" or

'universes' and try to grasp the immensity of just the physical part, alone, of all that 'down here'. Then strive to soar upwards into the *Non-Material Realms* of the *Eternal Part* of Creation, to *one* of the "many mansions in His Father's house"; that of our *true home* in The Spiritual Realm.
From "there" look downwards and strive to picture the **"7 Churches"** or 'universes' rotating like a huge wreath very, very far below.

If the *spirit* is *truly awake* to an experiential-recognition of *the moment*, the *mind-numbing revelation* becomes a ***spiritually-cathartic and life-changing event*** for the individual.

The final word on this subject must go to a woman who lectured in astronomy at Auckland City Observatory in New Zealand, and who was at the same time also a devout Christian. Upon enquiring why I sought the 1983 edition of the National Geographic Star Chart, I asked how she reconciled the time-frame that cosmology accepts for the age of the universe with that of 6,000 years which many millions of Christians faithfully but *mistakenly* believe. She replied that *that* was not a contentious issue for her.

She again asked why I wanted the 'Star Chart'; the question of which I perceived to be a very 'leading' one. Intuitively, therefore, I asked her if she knew of the "7 Churches" in **The Book of Revelation**, which she did. When made aware of the ***true nature*** of them, she asked – literally through a mask of what can only be described as something akin to "jaw-dropping, incredulity":

"Is that possible? Can it be possible?"

I stated to her that it was the only explanation which made logical sense [according to Inviolable and Immutable **Creation-Law**] and was therefore the ***actual reality***. As both a lecturer in astronomy and a Christian with some knowledge of The Bible, she *immediately understood* the ***full***

import of such a concept. A short while later she said –
clearly after some very serious reflection, however:

*"Thank you for that. I've always
believed we've tried to make GOD too
small."*

References

0.6 Bibliography

1. *The Holy Bible in Modern English*, Ferrar Fenton, Destiny Publishers, Massachusetts U.S.A. 1966 Edition.

2. *The Holy Bible, Authorised (King James) Version*, Eyre and Spottiswoode (Publishers) Ltd., Great Britain.

3. *A Gate Opens*, Herbert Vollmann, Composite Volume 1985, Stiftung Gralsbotschaft Publishing Co., Stuttgart.

4. *The Gospel of the Essenes*, The original Hebrew and Aramaic texts translated and edited by Edmund Bordeaux Szekely, Revised Edition, London, C. W. Daniel, 1976.

5. *Great Illustrated Dictionary Vol's I & II*, Reader's Digest, First Edition, 1984, USA.

6. *Ideas and Opinions*, Albert Einstein, Bonanza Books, New York, 1954.

7. *The Atlas of the Universe*, Patrick Moore, Mitchell Beazley Publishers, London, 1981 Edition.

8. *Astronomy Magazine* Edition April, 2009. Kalmbach Publishing Coy., Waukesha, Wi. USA.

9. *Seven Wonders of the Cosmos* Jayant V. Narlikar. Cambridge University Press, 1999.

10. *Telegraph Group Ltd.*

11. *The Sacred Balance: Rediscovering our place in nature,* David Suzuki, Allen and Unwin, Australia, 1997.

12. *Time Magazine 25th June 2001 Edition* (Feature Article: "How The Universe Will End")

13. *Reader's Digest Magazine, Aug 2004,* (Feature Article: "How to Make a Universe", Bill Bryson)

14. *National Geographic Star Chart,* 1983.

15. Brittanica CD '97.

0.7   The Parent Book:

BIBLE "MYSTERIES" EXPLAINED
[Revised Second Edition]
Understanding "Global Societal Collapse" from The "Science" in The Bible;
What Every Scientist, Bible Scholar and Ordinary Man Needs to Know!

The Revised Second Edition of 'Bible Mysteries...' is more comprehensive in that it now explains How and Why the 2008 global economic collapse occurred, but also When the seeds that wrought the How and Why were sown, and by Whom. [Chapter 3: **The Spiritual Laws: The Necessary Knowledge**
3.3.3 "Ten Men Will Take Counsel And It Will Come To Nought."
3.3.4 The Interlinked Global Monetary System "Reaping The Whirlwind." A Brief History Lesson.]

Additional information about the events surrounding the last day of Jesus's life, from His arrest in the Garden of Gethsemane to His murder at Golgotha, is now included. Using modern-day medical forensics, key experts – including two trauma surgeons – detail a harrowing blow-by-blow account of the terrible suffering of Jesus at the hands of His torturers; to His grievous 'walk' to Golgotha, and thence to His execution after the final excruciating hours on the death-cross.

The interesting question of the "Seven Churches in Asia-Minor" from The Book Of Revelation is examined more critically. Necessarily using the discoveries and mathematics of present-day astronomy/cosmology, the revealing conclusion of the true meaning perfectly resonates with the intuitive perception of the great mathematician, astronomer, theologian and scientist, Sir Isaac Newton.

This book, the result of many years of inner seeking and empirical research, offers *serious* seekers of the Truth a comprehensive understanding of the origin, meaning and purpose of human life; material and spiritual.

Beginning with **The Crucial Imperatives**: **Nine key points** that *must* be taken into consideration if logical and reasoned answers to humankind's Whence, Whither and Why is *ever* to be understood; the book takes the reader step by step through an understanding of man's **Spiritual Origins**, **The Spiritual Laws of Creation** [with sub-Chapter explaining the why of the universal pain of childbirth for human mothers; and the scientific quandary of the why of the enlarging baby cranium], the difference between **The First Death** and **The Second Death**; **Elemental Lore** [of Nature]; **Jesus! His Birth, Death and Resurrection** [a revisionist analysis]; before examining the truly 'mind-expanding' meaning of **"The 7 Churches in Asia"** from **The Book of Revelation**.

The key knowledge helps explain *why* there actually are **Two Sons of God** – final Chapter. It is key precisely because all other knowledge stems from that reality.

On reading the Work, the genuine seeker will clearly see that a conditioning process, set in place by religious authorities from the outset, over millennia has wrought appalling suffering through their inexcusable distortions of the Teachings of **The Truth** that once issued pristine and sublime from the Pure Holiness of its Bringer: **Jesus, The Son Of God!**

Now, because of those distortions, humankind is as a rudderless wreck on an increasingly stormy sea. Our many and increasing problems were not brought upon us by any kind of arbitrary randomness, but through *our constant and stubborn refusal to live according to the very Laws of Life which **alone** guarantee knowledge, peace and harmony.*

At the same time, however, – and precisely through the knowledge of those Laws – the way is shown in **how** we can *change* global societies *for the better.* Quite logically, if we continue down our present path for much longer *without such change*, the immutable outworking of **The Law** *will simply bring to an end* all that which *human thought and endeavour* had sought to establish and/or erect *in place of* the immutable and inviolable aegis of: **The One Law!** — — —

CREATION-LAW!

The Parent Work explains the How, the What and the Why!

Available in **U.S.A.: http://www.crystalbooks.org**

Or – **N.Z.:http://www.publishme.co.nz**

Table of Contents

4 Elemental Lore Of Nature

5 JESUS: His Birth, Death and Resurrection

6 Stigmata

7 Right Bible/Wrong Bible

8 The Emergence of Language

9 The First Death

0.8 The Booklet Series

*　　　*　　　*　　　*　　　*

THE TWO SONS OF GOD

The Son of Man and The Son of God
What The Bible Really Says

*　　　*　　　*　　　*　　　*

JESUS!:
His Birth, Death and
Resurrection

A Revisionist Analysis of the "Sacrosanct"
Christian Viewpoint

*　　　*　　　*　　　*　　　*

THE SPIRITUAL LAWS OF CREATION

The Crucial Knowledge for Humankind

*　　　*　　　*　　　*　　　*

WHITHER COMETH HUMANKIND?

(The Origins of Man) *Genesis and Science Agree*

*　　　*　　　*　　　*　　　*

THE "7 CHURCHES" Of THE
"REVELATION"

What the "Hubble" Will Never See
Sir Isaac Newton's "Plan of The World"

*　　　*　　　*　　　*　　　*

www.ingramcontent.com/pod-product-compliance
Lightning Source LLC
Chambersburg PA
CBHW071821020426
42331CB00007B/1573